METHODS IN MOLECULAR BIOLOGY

For further volumes:
http://www.springer.com/series/7651

For over 35 years, biological scientists have come to rely on the research protocols and methodologies in the critically acclaimed *Methods in Molecular Biology* series. The series was the first to introduce the step-by-step protocols approach that has become the standard in all biomedical protocol publishing. Each protocol is provided in readily-reproducible step-by step fashion, opening with an introductory overview, a list of the materials and reagents needed to complete the experiment, and followed by a detailed procedure that is supported with a helpful notes section offering tips and tricks of the trade as well as troubleshooting advice. These hallmark features were introduced by series editor Dr. John Walker and constitute the key ingredient in each and every volume of the *Methods in Molecular Biology* series. Tested and trusted, comprehensive and reliable, all protocols from the series are indexed in PubMed.

Double-Stranded RNA

Methods and Protocols

Edited by

Xiaofei Cheng

College of Plant Protection/Key Laboratory of Germplasm Enhancement, Physiology and Ecology of Food Crops in Cold Region of Chinese Education Ministry, Northeast Agricultural University, Harbin, China

Guanwei Wu

Institute of Plant Virology, Ningbo University, Ningbo City, Zhejiang, China

Humana Press

Editors
Xiaofei Cheng
College of Plant Protection/Key Laboratory of Germplasm Enhancement
Physiology and Ecology of Food Crops in Cold Region of Chinese Education Ministry
Northeast Agricultural University
Harbin, China

Guanwei Wu
Institute of Plant Virology
Ningbo University
Ningbo City, Zhejiang, China

ISSN 1064-3745 ISSN 1940-6029 (electronic)
Methods in Molecular Biology
ISBN 978-1-0716-3701-2 ISBN 978-1-0716-3702-9 (eBook)
https://doi.org/10.1007/978-1-0716-3702-9

This Humana imprint is published by the registered company Springer Science+Business Media, LLC, part of Springer Nature.
The registered company address is: 1 New York Plaza, New York, NY 10004, U.S.A.

Paper in this product is recyclable.

Preface

Double-stranded RNA (dsRNA) consists of two strands of RNA that are complementary to each other, forming a double-helix structure similar to DNA. It is the genetic material of some RNA viruses and the replication mediate of all RNA viruses. As a result, the presence of dsRNA is recognized as the hallmark of viral infection, and nearly all organisms have the capability of perception of dsRNA and mounting a response. For instance, dsRNA is recognized by Dicer or Dicer-like enzymes to trigger RNA interference (RNAi) in eukaryotic cells. In recent years, dsRNA-based approaches have emerged as a powerful tool for functional genomics research and have shown promise in combating viral infections and controlling pest populations in agriculture. As a result, dsRNA has gained significant attention in fields including plant biology, molecular biology, virology, entomology, and agriculture.

This volume is aimed to at providing detailed protocols concerning dsRNA. The volume consists of 18 protocols including classical and cutting-edge methods covering dsRNA isolation, visualization, characterization, production, and application. We tried to cover as many methods as possible for authors to select from. For instance, lithium chloride fractionation, cellulose chromatography, and a micro-spin column approach were characterized for dsRNA isolation in this volume. Each protocol is described in a step-by-step fashion to easily repeat and improve the protocol. Moreover, many protocols, such as co-immunoprecipitation-based isolation of double-stranded RNA-associated protein complexes, identification of mycoviruses by dsRNA extraction, application of dsRNA for fungi disease management (*Sclerotinia sclerotiorum* and *Botrytis cin*), and production of double-stranded RNA in plants by plant viral vectors for gene silencing, can also be easily adapted for identification of viruses from other organisms, control of other pathogens, and fundamental research.

We are grateful to all the authors for sharing their valuable working protocols. We apologize to the authors whose relevant work has not been cited and to readers for any missing dsRNA-related methods. Finally, we hope this volume will be a good reference book for students and researchers who work on dsRNA.

Harbin, China — *Xiaofei Cheng*
Ningbo City, Zhejiang, China — *Guanwei Wu*

Contents

Contributors

KEKELY BRUNO ATTIOGBE • *College of Plant Protection, Northeast Agricultural University, Harbin, China*
YANJU BAI • *Potato Research Institute of Heilongjiang Academy of Agricultural Sciences, Harbin, Heilongjiang, China*
MENGZHU CHAI • *College of Plant Protection, Northeast Agricultural University, Harbin, Heilongjiang, China; Key Laboratory of Germplasm Enhancement, Physiology and Ecology of Food Crops in Cold Region of Chinese Education Ministry, Northeast Agricultural University, Harbin, Heilongjiang, China*
SHUAIFA CHANG • *College of Plant Protection, Northeast Agricultural University, Harbin, Heilongjiang, China*
JIANPING CHEN • *State Key Laboratory for Managing Biotic and Chemical Threats to the Quality and Safety of Agro-products, Institute of Plant Virology, Ningbo University, Ningbo, China*
XIAOFEI CHENG • *College of Plant Protection, Northeast Agricultural University, Harbin, Heilongjiang, China; Key Laboratory of Germplasm Enhancement, Physiology and Ecology of Food Crops in Cold Region of Chinese Education Ministry, Northeast Agricultural University, Harbin, Heilongjiang, China*
SAIYA DUAN • *State Key Laboratory for Biology of Plant Diseases and Insect Pests, Institute of Plant Protection, Chinese Academy of Agricultural Sciences, Beijing, China; Scientific Observation and Experimental Station of Pests in Xilin Gol Rangeland, Ministry of Agriculture and Rural Affairs, Xilinhot, China*
WENQIAN FAN • *College of Plant Protection, Northeast Agricultural University, Harbin, Heilongjiang, China*
XINYUE FAN • *College of Plant Protection, Northeast Agricultural University, Harbin, Heilongjiang, China*
XINXIN FANG • *State Key Laboratory for Managing Biotic and Chemical Threats to the Quality and Safety of Agro-products, Institute of Plant Virology, Ningbo University, Ningbo, China*
YANLIN GAO • *Potato Research Institute of Heilongjiang Academy of Agricultural Sciences, Harbin, Heilongjiang, China*
RUOBING GUAN • *Henan International Laboratory for Green Pest Control, Henan Engineering Laboratory of Pest Biological Control, College of Plant Protection, Henan Agricultural University, Zhengzhou, China*
KAI GUO • *State Key Laboratory of Subtropical Silviculture, College of Forestry and Biotechnology, Zhejiang A & F University, Hangzhou, China*
LIHUA GUO • *State Key Laboratory for Biology of Plant Diseases and Insect Pests, Institute of Plant Protection, Chinese Academy of Agricultural Sciences, Beijing, China*
YATING GUO • *College of Plant Protection, Northeast Agricultural University, Harbin, China*
JUAN HUANG • *State Key Laboratory of Integrated Management of Pest Insects and Rodents, Institute of Zoology, Chinese Academy of Sciences, Beijing, China*

ZHAOXING JIA • *State Key Laboratory for Managing Biotic and Chemical Threats to the Quality and Safety of Agro-products, Institute of Plant Virology, Ningbo University, Ningbo, China*

XUE JIANG • *College of Plant Protection, Northeast Agricultural University, Harbin, China*

XAYVANGYE KORXEELOR • *College of Plant Protection, Northeast Agricultural University, Harbin, Heilongjiang, China*

YE LIU • *College of Plant Protection, Northeast Agricultural University, Harbin, Heilongjiang, China; Potato Research Institute of Heilongjiang Academy of Agricultural Sciences, Harbin, Heilongjiang, China*

HAICHAO LI • *Key Laboratory of Insect Developmental and Evolutionary Biology, CAS Center for Excellence in Molecular Plant Sciences, Chinese Academy of Sciences, Shanghai, China*

YAMENG LUAN • *College of Plant Protection, Northeast Agricultural University, Harbin, Heilongjiang, China*

TINGSHUAI MA • *College of Plant Protection, Northeast Agricultural University, Harbin, Heilongjiang, China*

XINYUE MA • *College of Plant Protection, Northeast Agricultural University, Harbin, Heilongjiang, China*

XUEXIA MIAO • *Key Laboratory of Insect Developmental and Evolutionary Biology, CAS Center for Excellence in Molecular Plant Sciences, Chinese Academy of Sciences, Shanghai, China*

DONGDONG NIU • *Department of Plant Pathology, College of Plant Protection, Nanjing Agricultural University, Nanjing, China; Key Laboratory of Integrated Management of Crop Diseases and Pests, Ministry of Education, Nanjing, China*

CLEMENT NZABANITA • *State Key Laboratory for Biology of Plant Diseases and Insect Pests, Institute of Plant Protection, Chinese Academy of Agricultural Sciences, Beijing, China*

PENGHUAN RUI • *State Key Laboratory for Managing Biotic and Chemical Threats to the Quality and Safety of Agro-products, Institute of Plant Virology, Ningbo University, Ningbo, China*

TAO TANG • *Department of Plant Pathology, College of Plant Protection, Nanjing Agricultural University, Nanjing, China; Key Laboratory of Integrated Management of Crop Diseases and Pests, Ministry of Education, Nanjing, China*

CHUNYU WANG • *State Key Laboratory of Subtropical Silviculture, College of Forestry and Biotechnology, Zhejiang A & F University, Hangzhou, China*

GUANGJUN WANG • *State Key Laboratory for Biology of Plant Diseases and Insect Pests, Institute of Plant Protection, Chinese Academy of Agricultural Sciences, Beijing, China; Scientific Observation and Experimental Station of Pests in Xilin Gol Rangeland, Ministry of Agriculture and Rural Affairs, Xilinhot, China*

YANFEI WANG • *State Key Laboratory for Biology of Plant Diseases and Insect Pests, Institute of Plant Protection, Chinese Academy of Agricultural Sciences, Beijing, China*

YUMENG WANG • *Department of Plant Pathology, College of Plant Protection, Nanjing Agricultural University, Nanjing, China; Key Laboratory of Integrated Management of Crop Diseases and Pests, Ministry of Education, Nanjing, China*

YUTING WANG • *College of Plant Protection, Northeast Agricultural University, Harbin, Heilongjiang, China*

ZIYI WANG • *College of Plant Protection, Northeast Agricultural University, Harbin, Heilongjiang, China*

GUANWEI WU • *State Key Laboratory for Managing Biotic and Chemical Threats to the Quality and Safety of Agro-products, Institute of Plant Virology, Ningbo University, Ningbo, China*
XIAOYUN WU • *College of Plant Protection, Northeast Agricultural University, Harbin, Heilongjiang, China*
KAI XU • *Jiangsu Key Laboratory for Microbes and Functional Genomics, Jiangsu Engineering and Technology Research Center for Microbiology, College of Life Sciences, Nanjing Normal University, Nanjing, China*
FEI YAN • *State Key Laboratory for Managing Biotic and Chemical Threats to the Quality and Safety of Agro-products, Institute of Plant Virology, Ningbo University, Ningbo, China*
JINGUANG YANG • *Key Laboratory of Tobacco Pest Monitoring, Controlling & Integrated Management, Tobacco Research Institute of Chinese Academy of Agricultural Sciences, Qingdao, China*
XIANGUANG YANG • *State Key Laboratory Base of Cell Differentiation and Regulation, College of Life Science, Henan Normal University, Xinxiang, China*
XIAOMING ZHANG • *State Key Laboratory of Integrated Management of Pest Insects and Rodents, Institute of Zoology, Chinese Academy of Sciences, Beijing, China; CAS Center for Excellence in Biotic Interactions, University of Chinese Academy of Sciences, Beijing, China; HainanYazhou Bay Seed Lab, Sanya, China*
YING ZHANG • *College of Plant Protection, Northeast Agricultural University, Harbin, Heilongjiang, China*
YUANMING ZHANG • *Jiangsu Key Laboratory for Microbes and Functional Genomics, Jiangsu Engineering and Technology Research Center for Microbiology, College of Life Sciences, Nanjing Normal University, Nanjing, China*
YU ZHAO • *College of Plant Protection, Northeast Agricultural University, Harbin, Heilongjiang, China*
WENLING ZHOU • *State Key Laboratory of Integrated Management of Pest Insects and Rodents, Institute of Zoology, Chinese Academy of Sciences, Beijing, China; CAS Center for Excellence in Biotic Interactions, University of Chinese Academy of Sciences, Beijing, China*

Chapter 1

Isolation of Double-Stranded RNAs by Lithium Chloride Fractionation

Mengzhu Chai, Xinyue Ma, Shuaifa Chang, and Xiaofei Cheng

Abstract

This procedure provides a comprehensive method for isolating double-stranded RNA (dsRNA) that relies on the different solubility of various nucleic acids in lithium chloride (LiC1). The approach offers several notable advantages including simplicity, avoidance of enzymatic treatments, and the ability to obtain relatively high yields of undegraded dsRNA over other conventional techniques. Moreover, it allows for the separation of different groups of cellular and viral nucleic acids from a single tissue sample. This method was further improved to increase the purity of dsRNA using plant tissues infected by RNA viruses.

Key words Double-stranded RNA, Lithium chloride, Fractionation, Centrifugation, Virus detection

1 Introduction

Extraction of double-stranded RNA (dsRNA) is a critical first step in virus detection. Lithium chloride (LiCl) precipitation is a well-established method for extracting and purifying RNA from varied tissues including plant cells [1]. It offers several advantages such as selectivity for RNA over DNA and proteins and the ability to denature RNases that could degrade the RNA. There has been a growing interest in improving the efficiency and purity of RNA extractions, especially in the plant virology research field [2–4]. This is because plant samples often contain high levels of secondary metabolites that can interfere with RNA quality. The LiCl fractionation procedure was firstly proposed by Diaz-Ruiz and Kaper in 1978 and was remained unchanged for decades [1]. Recently, Henderson demonstrated a method to enhance the LiCl fractionation procedure by incorporating an additional washing step with longer incubation times, which resulted in increased RNA yield and purity [5]. Ananda and colleagues further developed a modified LiCl protocol specifically for RNA extraction from plant tissues that are rich in polysaccharides and polyphenols [6]. These advancements in

Xiaofei Cheng and Guanwei Wu (eds.), *Double-Stranded RNA: Methods and Protocols*, Methods in Molecular Biology, vol. 2771,
https://doi.org/10.1007/978-1-0716-3702-9_1,

the LiCl-mediated RNA extraction methodology not only contribute to increasing the quality of the RNA yield but also expand the range of plant tissues from which high-quality RNA can be extracted. This chapter provides a comprehensive LiCl fractionation procedure for extracting dsRNAs from plant tissue infected by several well-known plant RNA viruses. Nevertheless, this method may be also useful for extracting dsRNAs from bacteria and animal tissue for diagnosing bacterial and animal viruses as well.

2 Materials

2.1 Chemical Reagents

All following chemicals are analytical grade. The manufacturers of the chemicals listed below are for reference only, and similar products from other companies can achieve the same effect. Store all chemicals at room temperature (unless indicated otherwise).

1. Liquid nitrogen.
2. Na-bentonite (e.g., Thermo Fisher Scientific, Catalog No. AAA1579536).
3. Glycine (e.g., Thermo Fisher Scientific, Catalog No. AA3643530).
4. Ethylenediaminetetraacetic acid (EDTA; e.g., Thermo Fisher Scientific, Catalog No. AAA1071336).
5. Sodium chloride (NaCl; e.g., Thermo Fisher Scientific, Catalog No. S271-1).
6. Sucrose (e.g., Thermo Fisher Scientific, Catalog No. S5-500).
7. Magnesium chloride hexahydrate ($MgCl_2{\cdot}6H_2O$; e.g., Sigma Aldrich, Catalog No. M9272).
8. Potassium chloride (KCl; e.g., Sigma Aldrich, Catalog No. P9541).
9. Sodium dodecyl sulfate (SDS; e.g., Sigma Aldrich, Catalog No. 436143).
10. Water-saturated phenol pH 6.6 (AquaPhenol; e.g., Thermo Fisher Scientific, Catalog No. 11AQUAPH01).
11. 8-hydroxyquinoline (e.g., Thermo Fisher Scientific, Catalog No. O261-100).
12. Anhydrous chloroform (e.g., MilliporeSigma, Catalog No. 50-947-683).
13. Anhydrous ethanol (e.g., Thermo Fisher Scientific, Catalog No. A405P-4).
14. Lithium chloride (LiCl; e.g., Thermo Fisher Scientific, Catalog No. AC449045000).

2.2 Extraction Buffer

Prepare the following solutions using ultrapure water and store all reagents at room temperature (unless indicated otherwise). Diligently follow all waste disposal regulations when disposing waste materials. All three buffers can be used to extract total nucleic acid and can be selected appropriately.

1. Nucleic acid extraction buffer A: 0.1 M glycine, 0.01 EDTA, 0.1 M NaCl, pH 9.5, 10% SDS, and 5% Na bentonite [1].
2. Nucleic acid extraction buffer B: 0.1 M EDTA, 0.1 M Tris-HCl, pH 8.9, 5% SDS, 40 mg/L Na bentonite [7].
3. Nucleic acid extraction buffer C: 0.2 M Tris-HCl, 0.4 M KCl, 0.2 M sucrose, 35 mM $MgCl_2$, 25 mM EDTA, pH 9.0 [8].

2.3 Equipment

1. Mortar and pestle.
2. Refrigerated Benchtop centrifuge (e.g., MIKRO200R, Germany).
3. Weighting scale.
4. Pipettes (20, 100, and 1000 μL).
5. Biosafety cabinet.
6. 15 and 50 mL falcon tubes.
7. 50 mL centrifuge tubes.
8. NanoDrop UV/Vis Spectrophotometer (e.g., NanoDrop 8000, Thermo Scientific).
9. Disposable Polypropylene micro-centrifuge tubes.
10. Water Baths (65 °C and 37 °C).
11. Gel electrophoresis system (e.g., Jordan Scientific JP-250).

3 Methods

3.1 Extraction of Total Nucleic Acids

1. Pre-chill the mortar and pestle with liquid nitrogen (*see* **Note 1**).
2. Homogenize fresh or frozen virus-infected leaf tissues (50–100 g) in liquid nitrogen with a chilled mortar with a pestle into fine powder (*see* **Note 2**).
3. Mix the leaf powder with prechilled nucleic acid extraction buffers at a ratio of 1 g per 1 mL buffer (mass/volume).
4. Transfer the homogenate to 250 mL centrifuge bottle after adding 200 mL of water-saturated phenol containing 0.1% 8-hydroxyquinoline.
5. Shake vigorously for 2 min and then centrifuge at 6500 rpm for 10 min at 4 °C.
6. Carefully transfer the upper aqueous phase to a new tube (*see* **Note 3**).

7. Repeat **steps 4–6** twice until no white interface substance is visible.
8. Add two volumes of cold 95% ethanol to the aqueous phase.
9. Mix gently and then incubate at −20 °C for 2 h.
10. Centrifugation at 6500 rpm for 10 min at 4 °C.
11. Discard the supernatant (*see* **Note 4**).
12. Dissolve the pellet in distilled water (total nucleic acid solution).

3.2 Extract Viral dsRNA by LiCl Fractionation

1. Add 8 M LiCl to the total nucleic acid solution to the final concentration of 2 M (*see* **Note 5**).
2. Incubate on ice for 8–18 h and then centrifuge at 10,000 rpm for 15 min at 4 °C.
3. Transfer the supernatant to a new test tube and add about 1/2 volume of 8 M LiCl to a final concentration of 4 M LiCl (*see* **Note 6**).
4. Incubate on ice for 8–18 h and then centrifuge at 10,000 rpm for 15 min at 4 °C.
5. Discard the supernatant and resuspend the pellets in ~50 μL distilled water (*see* **Note 7**).
6. Analyze the dsRNA by gel electrophoresis, as in Fig. 1, and/or microvolume spectrophotometer (*see* **Note 8**).
7. Store the dsRNA fraction at −80 °C for future use.

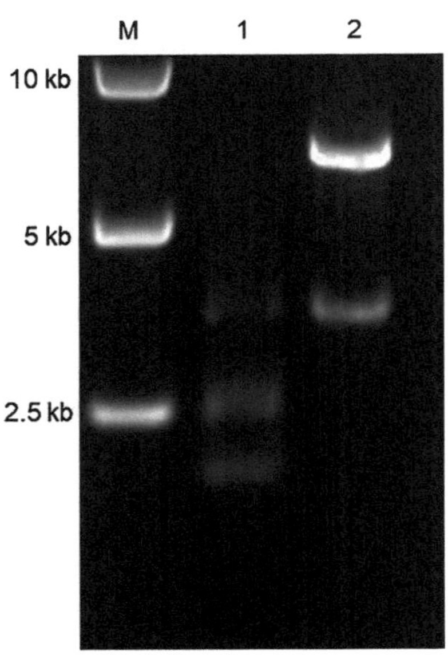

Fig. 1 dsRNA profiles in 1% Agar gel. M, nucleic acid marker; Lane 1 and 2 showing the dsRNA extracted from two field samples showing viral-like symptoms by lithium chloride fractionation

4 Notes

1. Always wear disposable gloves and change frequently during the operation to prevent the contamination of your samples by bacteria, fungi, and RNase present on hands.
2. Make sure the leaf tissue is homogenized into fine powder for maximum dsRNA yield.
3. Make sure to get rid of all the white interface substance and the bottom phenol phase to avoid the contamination of proteins and acquire maximum dsRNA yield.
4. Make sure the pellet is retained.
5. LiCl precipitation requires high-quality RNA since the contaminations in low-quality RNAs may lead to degradation and even complete loss of dsRNA during the experiment; SDS, urea, and diatomaceous earth have certain inhibitory effects on RNases and can be used interchangeably with each other during the precipitation.
6. The pellet contains high-molecular-weight single-stranded RNAs (ssRNAs), which can be collected if needed.
7. The supernatant contains DNA and soluble low-molecular-weight ssRNAs, which can be collected if needed.
8. Other nucleic acid analysis tools can also be used to estimate the quality of dsRNAs.

References

1. Diaz-Ruiz JR, Kaper JM (1978) Isolation of viral double-stranded RNAs using a LiCl fractionation procedure. Prep Biochem 8(1):1–17
2. Lot H, Kaper JM (1976) Physical and chemical differentiation of three strains of cucumber mosaic virus and peanut stunt virus. Virology 74(1):209–222
3. Kaper JM, Siberg RA (1969) Degradation of turnip yellow mosaic virus by freezing and thawing in vitro: a new method for studies on the internal organization of the viral components and for isolating native RNA. Virology 38(3):407–413
4. Morris TJ, Dodds JA (1979) Isolation and analysis of double-stranded RNA from virus-infected plant and fungal tissue. Phytopathology 69(8):854–858
5. Henderson JM, Ujita A, Hill E et al (2021) Cap 1 messenger RNA synthesis with co-transcriptional cleancap® analog by in vitro transcription. Curr Protocols 1(2):e39
6. Pratama AN, Grandmottet F, Rod-In W et al (2021) Efficient mini-prep RNA extraction from Dendrobium floral tissues rich in polysaccharides for validation of reference genes during flower development. Agric Natural Resour 55(3):349–358
7. Semancik JS, Weathers LG (1972) Exocortis virus: an infectious free-nucleic acid plant virus with unusual properties. Virology 47(2):456–466
8. Deng F, Xu R, Boland GJ (2003) Hypovirulence-associated double-stranded RNA from Sclerotinia homoeocarpa is conspecific with Ophiostoma novo-ulmi mitovirus 3a-Ld. Phytopathology 93(11):1407–1414

Chapter 2

Detection of dsRNA by Acridine Orange Staining

Tingshuai Ma, Yu Zhao, and Xiaofei Cheng

Abstract

Acridine orange is a nucleic acid binding dye that emits green fluorescence when bound to double-stranded DNA or RNA and red fluorescence when bound to single-stranded DNA or RNA under ultraviolet light. This unique characterization allows it to be used for distinguishing or visualization of dsRNA. Here, we present a convenient and efficient protocol for detecting dsRNA in polyacrylamide gels.

Key words dsRNA, Acridine orange, In-gel staining

1 Introduction

Acridine orange (AO) is a fluorescent, nucleic acid intercalating dye that has been widely used in cytochemistry to track pH gradients in biological samples, nuclear fragmentations, flow cytometry, and other applications [1, 2]. AO was originally extracted from coal tar and was used as a dye and antibacterial agent in the fabric industry in the late nineteenth century [3, 4]. In 1940s, AO was discovered to have the capacity to bind nucleic acids [5]. Shortly, it was reported that AO has different affinities for double-stranded and single-stranded nucleic acids and emits different fluorescent signal after binding double-stranded and single-stranded acids: a green fluorescence at 530 nm when binds to double-stranded nucleic acids and red fluorescence at 640 nm when binds to single-stranded nucleic acids [6]. This unique property makes it an ideal agent for distinguishing double-stranded and single-stranded nucleic acids and stain cells [7, 8].

Double-stranded RNA (dsRNA) is the genetic material of dsRNA viruses and is produced during the replication of all RNA viruses. As a result, the presence of long dsRNA is recognized as a hallmark of virus infection. DsRNA is also crucial for numerous

Xiaofei Cheng and Guanwei Wu (eds.), *Double-Stranded RNA: Methods and Protocols*, Methods in Molecular Biology, vol. 2771, https://doi.org/10.1007/978-1-0716-3702-9_2,

biological functions of host cells such as triggering RNA silencing, signaling transporting, and processing of endogenous RNAs, e.g., microRNAs, transfer RNAs, and ribosome RNAs [9, 10]. AO emits red and green fluorescence when bound to single-stranded and double-stranded nucleic acids, respectively; while other nucleic acid dyes such as ethidium bromide (EB) lack color bias [11, 12]. This unique characteristic can be used to identify dsRNA from total RNAs and search for dsRNA from virus-infected cells [13, 14]. In this protocol, we provide a quick and reliable method for identifying dsRNA from total RNAs extracted from plant samples by AO staining in polyacrylamide gel.

2 Materials

2.1 Plant Materials and Chemicals

1. Liquid nitrogen.
2. Hexadecyltrimethyl ammonium bromide (CTAB).
3. Polyvinylpyrrolidone (PVP).
4. Dithiothreitol (DTT).
5. Chloroform (e.g., MilliporeSigma, Catalog No. 50-947-683).
6. Isoamyl alcohol.
7. Centrifuge tube.
8. Shaker (e.g., Thermomixer comfort, Eppendorf).
9. Lithium chloride (LiCl; e.g., Thermo Fisher Scientific, Catalog No. AC449045000).
10. Anhydrous ethanol (e.g., Thermo Fisher Scientific, Catalog No. A405P-4).
11. Tris-base.
12. Ethylenediaminetetraacetic acid (EDTA; e.g., Thermo Fisher Scientific, Catalog No. AAA1071336).
13. Sodium chloride (NaCl).
14. Sodium hydroxide (NaOH).
15. Beta-mercaptoethanol.
16. Diethyl pyrocarbonate (DEPC).
17. Acridine orange.
18. Agarose.
19. Acetate.
20. DNase I (e.g., Vazyme DNase I, RNase-free; cat no. EN401).
21. RNA loading dye (e.g., New England BioLabs 2×RNA loading dye; cat no. B0363S).

2.2 Equipment

1. Mortar and pestle.
2. Pipettes (20, 100, and 1000 μL).
3. Refrigerated Benchtop centrifuge (e.g., MIKRO200R, Germany).
4. Weighting scale.
5. Biosafety cabinet.
6. Tissuelyser (e.g., TissueLyser II – QIAGEN).
7. Electrophoresis device (including casting tray and well combs).
8. UV transilluminator.
9. Microwave.
10. Flask.
11. Water bath or thermal cycle.

2.3 Buffer Solutions (See Note 1)

1. CTAB buffer: 2.0% CTAB, 2.0% PVP 40, 1.4 M NaCl, 100 mM Tris-HCl (pH 8.0), 20 mM EDTA, and 1.0% beta-mercaptoethanol in RNase-free water.
2. CIA solution: mix 96 mL anhydrous chloroform and 4 mL isoamyl alcohol in biosafety hood at room temperature.
3. 8 M LiCl.
4. 1.0 M Tris-HCl (pH 8.0).
5. 0.5 M EDTA solution: weight 186.1 g EDTA and 18 g NaOH; stir the solution vigorously using a magnetic stirrer until EDTA dissolved completely; top up the solution to 1 L with DEPC-treated distilled water (*see* **Note 2**).
6. TE buffer: Measure out 1 mL 1M Tris-Cl (pH 8.0) and 0.2 mL 0.5 M EDTA (pH 8.0) to a 100 mL Duran bottle; Top up the solution to 100 mL by distilled RNase-free water.
7. 70% ethanol: Measure out 70 mL anhydrous ethanol and top up to 100 mL by distilled RNase-free water.
8. 50×TAE buffer: Weight 242 g Tris-base, measure out 57.1 mL 100% acetic acid, and 100 mL 0.5 M EDTA (pH 8.0); top up to 1 L by DEPC-treated distilled water.

3 Methods

3.1 Preparation of Total RNA

1. Preheat CTAB buffer to 65 °C before use.
2. Ground fresh leaf tissue into fine powder by a prechilled mortar and pestle or via a tissuelyser (*see* **Note 3**).
3. Transfer leaf powder to a 2 mL centrifuge tube containing 900 μL preheated CTAB buffer.
4. Gentle vortex for 10 s.

5. Incubate at 65 °C for 20 min on a shaker.
6. Centrifuge at 11,000 rpm for 10 min at room temperature.
7. Carefully transfer the supernatant to a new 2 mL centrifuge tube and add equal volume of CIA solution.
8. Vortex for 10 s by hand and centrifuge at 12,000 rpm for 10 min at room temperature.
9. Repeat **steps** 7 and **8** once.
10. Add prechilled 8 M LiCl to the final concentration of 4.0 M.
11. Mix gently and incubate at 4 °C overnight to precipitate RNA.
12. Centrifuge at 13,000 rpm for 20 min at 4 °C.
13. Discard the supernatant and add 500 μL prechilled 70% ethanol.
14. Mix gently and centrifuge at 13,000 rpm for 1 min at 4 °C.
15. Repeat **steps 13** and **14** once.
16. Discard the supernatant and add 500 μL prechilled 100% ethanol.
17. Mix gently and centrifuge at 13,000 rpm for 1 min at 4 °C.
18. Discard the supernatant and dissolve the RNA pellets in 100 μL TE buffer.

3.2 DNase Treatment

1. Measure 1 μg total RNA, add 1 μl 10×DNase reaction buffer, 1 μL DNase I and top up to 10 μL by DEPC-treated distilled water (*see* **Note 4**).
2. Mix well and incubate at 37 °C for 15 min (*see* **Note 5**).
3. Add 1 μL of 25 mM EDTA (pH 8.0) and mix well by pipetting.
4. Incubate at 65 °C for 15 min to heat inactivate the DNase I.
5. Place on ice for 1 min; this RNA solution can directly be used for subsequent gel electrophoresis and reverse transcription reaction.

3.3 Gel Electrophoresis and Acridine Orange Staining

1. Measure 1 g of agarose and add 100 mL 1×TAE in a microwavable flask.
2. Microwave for 1–3 min until the agarose is completely dissolved (*see* **Note 6**).
3. Let agarose solution cool down to about 50 °C (*see* **Note 7**).
4. Add acridine orange to a final concentration of approximately 0.2–0.5 μg/mL.
5. Pour the agarose into a gel tray with the well comb in place (*see* **Note 8**).
6. Cold down the gel at 4 °C for 10–15 min or at room temperature for 20–30 min.

7. Place the gel into the gel tank and fill the tank will 1 TAB until the gel is completely covered.
8. Add equal volume of 2×RNA loading dye to each sample, mix well, and load into gel well.
9. Carefully load an RNA molecular weight ladder into the first lane of the gel set the electrophoresis unit.
10. Set up the electrophoresis set and run the gel at 90 V until the bromophenol blue is two-thirds of the way to the bottom (*see* **Note 9**).
11. Visualize the gel on a UV transilluminator and then use Photoshop CS to eliminate the blue channel that depicts the UV lamp (*see* **Note 10**).

4 Notes

1. Prepare all buffers with DEPC-treated or RNase-free water.
2. EDTA hardly dissolves under acid condition; sodium hydroxide should be added gradually to increase the solubility.
3. Make sure to ground the tissue into fine powder to achieve maximum dsRNA yield.
4. The reaction can be scaled up based on the total amount of RNA.
5. DNase treatment is often incomplete at 25 °C. DNase I is more effective at 37 °C.
6. Pulses model and swirling the flask occasionally is preferred to avoid overboiling the solution.
7. About when you can hold the flask (takes about 5 min).
8. Pour slowly to avoid bubbles; use the well-comb pipette tip to eliminate bubbles or push them to the sides of the gel if necessary.
9. Usually takes approximately 60 min to finish the electrophoresis.
10. A chemiluminescent imaging system with an adjustable light spectrum is more desirable.

References

1. Palmgren MG (1991) Acridine orange as a probe for measuring pH gradients across membranes: mechanism and limitations. Anal Biochem 192(2):316–321
2. Hitoshi Y, Lorens J, Kitada SI, Fisher J, LaBarge M, Ring HZ, Francke U, Reed JC, Kinoshita S, Nolan GP (1998) Toso, a cell surface, specific regulator of Fas-induced apoptosis in T cells. Immunity 8(4):461–471
3. Wainwright M (2001) Acridine-a neglected antibacterial chromophore. J Antimicrob Chemother 47(1):1–13
4. Browning CH, Cohen JB, Gaunt R, Gulbransen R (1997) Relationships between antiseptic

action and chemical constitution with special reference to compounds of the pyridine, quinoline, acridine and phenazine series. Proc R Soc Lond Ser B 93(653):329–366

5. Von Bertalanffy L, Bickis I (1956) Identification of cytoplasmic basophilia (ribonucleic acid) by fluorescence microscopy. J Histochem Cytochem 4(5):481–493
6. McMaster GK, Carmichael GG (1977) Analysis of single- and double-stranded nucleic acids on polyacrylamide and agarose gels by using glyoxal and acridine orange. Proc Natl Acad Sci 74(11):4835–4838
7. Lillie RD, Conn HJ, Stotz EH, Emmel VM (1977) H. J. Conn's biological stains: a handbook on the nature and uses of the dyes employed in the biological laboratory, 9th edn. Williams and Wilkins, Baltimore
8. Plemel JR, Caprariello AV, Keough MB, Henry TJ, Tsutsui S, Chu TH, Schenk GJ, Klaver R, Yong VW, Stys PK (2017) Unique spectral signatures of the nucleic acid dye acridine orange can distinguish cell death by apoptosis and necroptosis. J Cell Biol 216(4):1163–1181
9. Jacobs BL, Langland JO (1996) When two strands are better than one: the mediators and modulators of the cellular responses to double-stranded RNA. Virology 219(2):339–349
10. Ablasser A, Bauernfeind F, Hartmann G, Latz E, Fitzgerald KA, Hornung V (2009) RIG-I-dependent sensing of poly(dA:dT) through the induction of an RNA polymerase III–transcribed RNA intermediate. Nat Immunol 10(10):1065–1072
11. Carmichael GG, McMaster GK (1980) The analysis of nucleic acids in gels using glyoxal and acridine orange. Methods Enzymol 65(1):380–391
12. Bruno JG, Sincock SA, Stopa PJ (1996) Highly selective acridine and ethidium staining of bacterial DNA and RNA. Biotech Histochem 71(3):130–136
13. Pichlmair A, Schulz O, Tan CP, Rehwinkel J, Kato H, Takeuchi O, Akira S, Way M, Schiavo G, Reis e Sousa C (2009) Activation of MDA5 requires higher-order RNA structures generated during virus infection. J Virol 83(20):10761–10769
14. Lauretti F, Lucas de Melo F, Benati FJ, de Mello Volotão E, Santos N, Linhares RE, Nozawa C (2003) Use of acridine orange staining for the detection of rotavirus RNA in polyacrylamide gels. J Virol Methods 114(1):29–35

Chapter 3

Isolation of dsRNA from Plants by Cellulose Chromatography

Yameng Luan, Wenqian Fan, Xayvangye Korxeelor, and Xiaoyun Wu

Abstract

As the constitutive molecules of double-stranded RNA (dsRNA) virus genomes and replicative intermediates of single-stranded RNA (ssRNA) viruses, the high-molecular-weight dsRNAs are commonly found in RNA virus-infected plants. Therefore, the dsRNA is recognized as a hallmark of RNA virus infection and the profile of dsRNA has been applied as an efficient tool for diagnoses or characterization of unreported RNA viruses. Cellulose chromatography is one of the most useful procedures for the isolation of viral dsRNAs from total nucleic acids. Here, we describe rapid cellulose-based methods for purification of dsRNAs from plant tissue.

Key words Double-stranded RNA, Isolation methods, Cellulose chromatography, Plant virus

1 Introduction

Varied serological or polymerase chain reaction (PCR)-based diagnosis methods such as enzyme-linked immunosorbent assay (ELISA), Western blot, dot blot, reverse transcription-PCR (RT-PCR), and recombinase polymerase amplification-lateral flow assays (RPA-LFA) have been developed for rapid and reliable detection of viral diseases [4]. These techniques require antigen or sequence information of target viruses; thus, are not suitable for the diagnosis/detection of unknown viruses. The majority of all currently known viruses, especially plant viruses, contain RNA genomes [5]. The high-molecular-weight double-stranded RNAs (dsRNA) are constitutive materials of dsRNA virus genomes and the replicative intermediates of all RNA viruses. Therefore, the presence of dsRNA has been recognized as a hallmark of RNA virus infection [3, 7, 8]. With stability and easy extraction, dsRNA profiling is an efficient tool for diagnoses or characterization of RNA viruses but not reliable for viruses with DNA genomes

Xiaofei Cheng and Guanwei Wu (eds.), *Double-Stranded RNA: Methods and Protocols*, Methods in Molecular Biology, vol. 2771, https://doi.org/10.1007/978-1-0716-3702-9_3,

[9]. Moreover, the extracted dsRNA can be used for further the identification of unknown RNA viruses through high-throughput sequencing.

Several methods have been developed for the extraction of dsRNA such as cellulose chromatographic separation, enzymatic degradation of ssRNA and DNA, and lithium chloride fractionation [1]. Cellulose column chromatography is one of the most useful procedures for isolating viral dsRNAs from total nucleic acids. Here, we describe a rapid dsRNA extraction method using CF-11 cellulose and free of phenol and chloroform.

2 Materials

2.1 Equipment

1. Mortar and pestle.
2. Funnel.
3. Precision analytical digital lab scale.
4. Refrigerated Centrifuge.
5. Pipettes.
6. Biosafety cabinet.
7. Centrifuge tubes.
8. NanoDrop UV/Vis Spectrophotometer (e.g., NanoDrop 8000, Thermo Scientific).
9. Gel electrophoresis system.

2.2 Chemical and Regents

1. Liquid nitrogen.
2. CF11 cellulose (Whatman).
3. Ethanol.
4. Polyvinylpolypyrrolidone (PVPP-40, Sigma-Aldrich).
5. Agarose.
6. Tris base.
7. EDTA.
8. Anhydrous ethanol.
9. RNase-free water.
10. RQ1 DNase (Promega).
11. Polyvinylpolypyrrolidone 40 (PVPP-40, Sigma-Aldrich).
12. NaCl.
13. DEPC.
14. SDS.
15. β-mercaptoethanol.
16. Sodium acetate.
17. Isopropanol.

2.3 Buffers

1. 10% SDS (*see* **Note 1**).
2. 1 M Tris-HCl (pH 8.0).
3. 0.5 M EDTA (pH8.0) (*see* **Note 2**).
4. 10% PVPP-40.
5. Diethyl pyrocarbonate (DEPC).
6. Nucleic acid extraction (NAE) buffer: 50 mM Tris-HCl (pH 8.5), 50 mM EDTA, 3% SDS, 1% β-mercaptoethanol, 1% PVPP-40.
7. 10×dsRNA-binding (10×DSB) buffer: 0.1 M Tris-HCl (pH 8.0), 1 M NaCl, 0.01 M EDTA.
8. 1×DSB buffer (containing 20% Ethanol): 10 mM Tris-HCl (pH 8.0), 100 mM NaCl, 1 mM EDTA, 20% ethanol.
9. Elution buffer: 10 mM Tris-HCl (pH 8.0), 100 mM NaCl, 1 mM EDTA.
10. 3 M sodium acetate (pH 5.2).

3 Methods

This protocol is based on the previous description with some modifications [2, 6]. Compared to phenol/chloroform extraction of dsRNA, this method is free of phenol and chloroform and is highly efficient and rapid (less than 2 h to complete whole steps and only requires about 200 mg of fresh material). All reagents and buffers should be prepared with DEPC-treated distilled water or directly with distilled RNase-free water (*see* **Note 3**).

1. Precool the mortar and pestle in liquid nitrogen.
2. Grind about 200 mg virus-infected plant sample in liquid nitrogen into fine powder (*see* **Note 4**).
3. Transfer the tissue powder to a 2 mL centrifuge tube containing 600 μL NAE buffer and add anhydrous ethanol to the final concentration of 30% (*see* **Note 5**).
4. Thaw the tissue powder by vortexing immediately and then shake the mixture gently for at least 20 min at room temperature on a horizontal shaker (*see* **Note 6**).
5. Centrifuge at 10,000× *g* for 20 min at 4 °C.
6. Equilibrate 40 mg cellulose CF-11 in 600 μL 1×DSB buffer in a centrifuge tube (*see* **Note 7**).
7. Collect cellulose CF-11 by centrifuge and discard the supernatant thoroughly.
8. Transfer the supernatant of **step 5** to the centrifuge tube containing equilibrated cellulose CF-11 from **step 7**.

9. Mix well by gently vortexing, shake the mixture gently for 30 min at room temperature on a horizontal shaker (*see* **Note 8**).
10. Centrifuge at 5000× *g* for 1 min at room temperature to collect the cellulose CF-11.
11. Remove the supernatant and add 1 mL of 1×DSB buffer.
12. Mix well with a pipette and then shake gently for 5 min on a horizontal shaker at room temperature.
13. Repeat **steps 10–12** twice (*see* **Note 9**).
14. Remove the supernatant completely by pipetting.
15. Add 200 μL elution buffer to the dried pellet, vortex gently for 5 min on a horizontal shaker, and then centrifuge at 5000× *g* for 1 min at room temperature.
16. Transfer the supernatant to a new 1.5 mL centrifuge tube kept on ice.
17. Repeat **step 15** once.
18. Centrifuge at 5000× *g* for 1 min at room temperature to collect the supernatant to the same tube of **step 16**.
19. Centrifuge the tube containing dsRNA for 1 min at 5000× *g* at 4 °C and transfer the supernatant to a new 1.5 mL tube to eliminate any remaining cellulose.
20. Add 1/10 volume of 3 M sodium acetate (pH 5.2) and 0.8 volume of isopropanol to the supernatant, mix well, and incubate overnight at −20 °C or 1 h at −80 °C.
21. Centrifuge at >14,000× *g*, 4 °C for 30 min to collect the dsRNA pellet.
22. Wash the dsRNA pellet with 70% ethanol twice by centrifugation.
23. Discard the supernatant, air dry the pellet in a biosafe cabinet for 10 min and dissolve the dsRNA pellet in about 200 μL DEPC-treated water (*see* **Note 10**).

4 Notes

1. SDS precipitates at low temperatures. The solution should be prepared with heating and needs to be warmed to at least room temperature prior to use.
2. EDTA is difficult to dissolve at high pH; therefore, it is necessary to measure the pH of the solution when adding HCl to adjust the pH.

3. Glass pipettes should be used to measure DEPC since some plastic pipettes can be dissolved by DEPC. DEPC treatment must be handled in a fume hood for safety. DEPC should be inactivated by autoclave at 121 °C for at least 15 min.
4. Use young and symptomatic leaf tissue and make sure to ground the leaf tissue into fine powder to maximize the dsRNA yield.
5. Use RNase-free pipette tips with a filter to avoid contamination among samples.
6. Avoid foam during vortex.
7. Equilibration can increase the binding activity of cellulose CF-11.
8. The incubation time can be increased.
9. The washing steps can be increased until the supernatant becomes colorless to eliminate the contamination of cytochrome.
10. The dsRNA solution may contain DNA and single-stranded RNAs, which can be eliminated by DNase and RNase treatment.

References

1. Atsumi G, Sekine K-T, Kobayashi K (2015) A new method to isolate total dsRNA. In: Uyeda I, Masuta C (eds) Plant virology protocols: new approaches to detect viruses and host responses. Springer, New York, pp 27–37
2. Balijja A, Kvarnhedenb A, Turchetti T (2008) A non-phenol-chloroform extraction of double-stranded RNA from plant and fungal tissues. Journal of Virological Methods 152:32–37
3. Dodds JA, Morris TJ, Jordan RL (1984) Plant viral double-stranded RNA. Annual Review of Phytopathology 22:151–168
4. Kalimuthu K, Arivalagan J, Mohan M, Christyraj JRSS, Arockiaraj J, Muthusamy R, Ju H-J (2022) Point of care diagnosis of plant virus: current trends and prospects. Molecular and Cellular Probes 61:101779
5. King AM, Lefkowitz E, Adams MJ, Carstens EB (2011) Virus taxonomy: ninth report of the International Committee on Taxonomy of Viruses. Elsevier
6. Marais A, Faure C, Bergey B, Candresse T (2018) Viral double-stranded RNAs (dsRNAs) from plants: alternative nucleic acid substrates for high-throughput sequencing. In: Pantaleo V, Chiumenti M (eds) Viral metagenomics: methods and protocols, vol 1746. Springer, New York, pp 45–53
7. Morris TJ, Dodds JA, Hillman B, Jordan RL, Lommel SA, Tamaki SJ (1983) Viral specific dsRNA: diagnostic value for plant virus disease identification. Plant Molecular Biology Reporter 1:27–30
8. Valverde R, Nameth S, Jordan R (1990) Analysis of double-stranded RNA for plant virus diagnosis. Plant Disease 74:255–258
9. Wu Q, Ding S-W, Zhang Y, Zhu S (2015) Identification of viruses and viroids by next-generation sequencing and homology-dependent and homology-independent algorithms. Annual Review ofPhytopathology 53: 425–444

Chapter 4

Rapid Purification of dsRNA Using Micro-Spin Cellulose Column

Yuting Wang, Ziyi Wang, and Xiaoyun Wu

Abstract

Double-stranded RNA (dsRNA) is the replication intermediates of all RNA viruses. Purification and analysis of the profile and sequence of dsRNA is vital in virus diagnoses and/or characterization. Cellulose is one of the common materials used for isolation of dsRNA. Cellulose specifically binds dsRNA fraction under 15% ethanol concentration, which allows to isolate dsRNA from total nucleic acid solution or cell lysate. Here, we describe a rapid and reliable method for purifying dsRNA using a home-made micro-spin cellulose column from the cell lysate of virus-infected plant tissue. This labor-saving and rapid method enables routinely high-throughput isolation and analysis of dsRNA in plant or fungi samples.

Key words Double-stranded RNA, Isolation methods, Cellulose chromatography, Plant virus

1 Introduction

Precise diagnosis of viruses mainly relies on serological or polymerase-based virus diagnosis methods that have been developed such as immunochromatographic strip, enzyme-linked immunosorbent assay (ELISA), Western blot, reverse transcription-polymerase chain reaction (RT-PCR), and recombinase polymerase amplification-lateral flow assays (RPA-LFA) [1]. However, these methods can only detect known viruses. Traditionally, diagnosis of unknown viruses mainly relies on electron microscope and the combination of those serological or polymerase-based methods and even the experience of the researcher. The recently developed next-generation high-throughput sequencing (NGS) allows to discover known and unknown viruses without bias. Although total cellular RNA can be used for constructing of sequencing library for NGS, the mast number of host rRNA, small RNA, and mRNA greatly reduced the sensitivity of this method. RNA is the genetic material of most plant, animal, fungal, and bacterial viruses [2]. The replication of all RNA viruses

Xiaofei Cheng and Guanwei Wu (eds.), *Double-Stranded RNA: Methods and Protocols*, Methods in Molecular Biology, vol. 2771, https://doi.org/10.1007/978-1-0716-3702-9_4,

must go through the dsRNA phase as all genetic information in RNA is copied through base pairing similar as the DNA counterpart [3–5]. Therefore, dsRNA is the ideal material for NGS analysis of known and unknown RNA viruses [6]. Cellulose is one of the most powerful materials for isolating dsRNAs as it specifically and reversibly binds to dsRNA [7–9]. Here, we report simple and reliable dsRNA extraction methods using a home-made microspin cellulose column modified from Okada et al. [10].

2 Materials

2.1 Equipment

1. Glass measuring cylinders.
2. Mortar and pestle or electronic homogenizer (e.g., Qiagen TissueLyser II).
3. 2.0 mL and 0.6 mL centrifuge tubes.
4. # 22-gauge syringe needle.
5. Analytical lab scale.
6. Pipettes.
7. Biosafety cabinet.
8. Gel electrophoresis system.

2.2 Chemical and Regents

1. Cellulose (Whatman CF-11 Cellulose powder; Cat. no. 4021-050; Merck Sigma-Aldrich Cellulose; Cat no. C6288).
2. Liquid nitrogen.
3. Chloroform: isoamyl alcohol, TE saturated (25:24:1).
4. Agarose.
5. Tris-base.
6. Ethylenediaminetetraacetic acid (EDTA).
7. SYBR green.
8. Anhydrous ethanol.
9. Polyvinylpolypyrrolidone 40 (PVPP-40, Sigma-Aldrich).
10. Sodium chloride (NaCl).
11. Diethyl pyrocarbonate (DEPC).
12. Sodium dodecyl sulfate (SDS).
13. Sodium acetate.
14. β-mercaptoethanol.
15. 2×RNA loading buffer.
16. Hydrogen chloride (HCl).
17. Sodium hydroxide (NaOH).

2.3 Buffers

1. DEPC-treated, autoclaved water: measure 1 mL of DEPC and top to 1 L with ultrapure water, shake vigorously and allow to stand overnight at room temperature, then autoclave at 121 °C for 20 min (*see* **Note 1**).
2. 10% SDS (*see* **Note 2**).
3. 1 M Tris-HCl (pH 8.0): weight 12.11 g Tris base and add 80 mL DEPC-treated, autoclaved water, adjust the pH to 8.0 with HCl, and then top up to 100 mL with DEPC-treated, autoclaved water.
4. 0.5 M EDTA (pH 8.0) solution: weight 18.61 g dihydrate EDTA disodium salt add 80 mL DEPC-treated, autoclaved water, adjust the pH to 8.0 with solid NaOH pellets or powder carefully; top up to 100 mL with DEPC-treated, autoclaved water.
5. 10×Tris-EDTA-Sodium chloride (TES) buffer: 0.1 M Tris-HCl (pH 8.0), 1 M NaCl, 0.01 M EDTA.
6. Washing buffer (100 mM NaCl; 10 mM Tris–HCl, pH 8.0; 1 mM EDTA, pH 8.0, 16% ethanol): Take 10 mL 10×TSE buffer and 16 mL anhydrous ethanol in a 100 mL glass measuring cylinder, and top up to 100 mL with DEPC-treated distilled water.
7. Extraction buffer [200 mM NaCl; 20 mM Tris–HCl, pH 8.0; 2 mM EDTA, pH 8.0; 1% (w/v) SDS; 0.1% (v/v) beta-mercaptoethanol]: Take 2 mL 10×TSE buffer, 1 mL 10% SDS, and 10 μL beta-mercaptoethanol in a 10 mL glass measuring cylinder, and top up to 10 mL with DEPC-treated distilled water.
8. 3 M Sodium acetate (pH 5.2): Add 40.8 g sodium acetate trihydrate to 80 mL of water. After dissolution, transfer the bottle to a fume hood and adjust the pH to 5.2 with pure (glacial) acetic acid. Make up to 100 mL with water and autoclave.
9. 33%, 75% ethanol solution: Measure 33 mL or 75 mL anhydrous ethanol and top up to 100 mL with DEPC-treated distilled water.

3 Methods

All reagents should be prepared with high-purity, DEPC-treated, autoclaved water. A brief schematic diagram of the entire procedure is illustrated in Fig. 1.

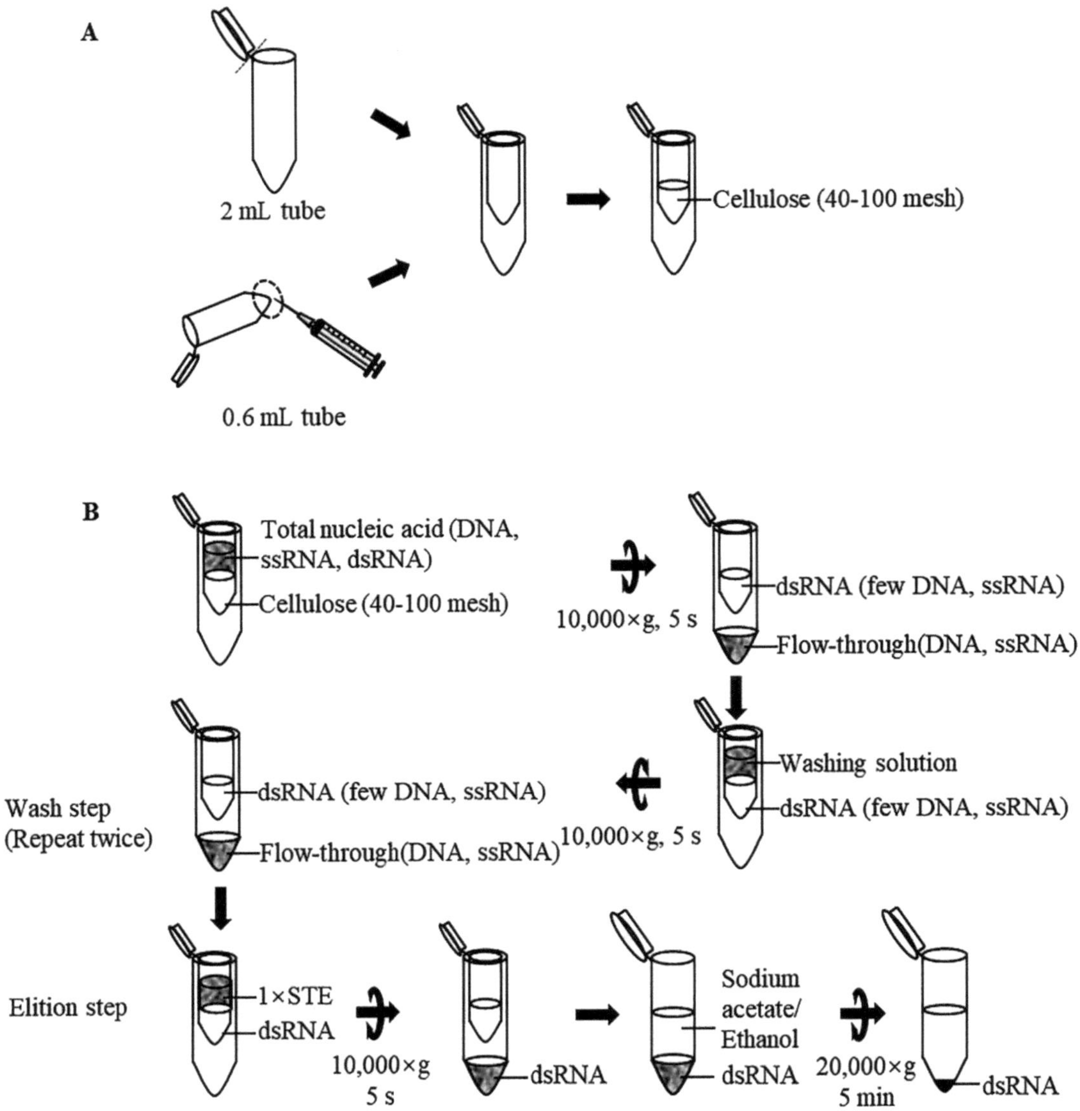

Fig. 1 Schematic representation of the dsRNA purification procedure. (**a**) Preparation of the micro-spin column device; (**b**) purification steps

3.1 Preparation of the Micro-spin Column Device

1. Take a 0.6 mL centrifuge tube and make one or two holes at the bottom with a syringe needle of outer diameter 0.7 mm or less (*see* **Note 3**).
2. Place the 0.6 mL tube in a 2.0 mL centrifuge tube with its cap removed.
3. Prepare cellulose slurry by mixing an equal volume of cellulose powder with washing buffer (*see* **Note 4**).
4. Pack the 0.6 mL tube with 600 μL cellulose slurry.
5. The micro-spin column device can be used immediately or stored at 4 °C until use (*see* **Note 5**).

3.2 Purification of dsRNA by Micro-spin Column

1. Pulverize fresh leaf tissue (about 200 mg) in liquid nitrogen with the precooled mortar and pestle or tissuelyser (*see* **Note 6**).
2. Carefully transfer the tissue powder to a 2.0 mL centrifuge tube that is prefilled with 500 μL extraction buffer and mix well by vertically flipping (*see* **Note 7**).
3. Add an equal volume of phenol:chloroform:isoamyl alcohol (25:24:1, v/v) to the crude extract.
4. Vortex the mixture for 1 min and centrifuge at 20,000× *g* for 5 min to clarify the solution.
5. Carefully move the supernatant to a new centrifuge tube by pipetting (*see* **Note 8**).
6. Repeat **steps 4** and **5** again if the supernatant remains cloudy (*see* **Note 9**).
7. Carefully transfer the supernatant to a new 1.5 mL tube and add an equal volume of 33% ethanol (final concentration of about 16.5%) (*see* **Note 10**).
8. Centrifuge the mixture at 20,000× *g* for 3 min to eliminate any precipitates.
9. Centrifuge the micro-spin column device 10,000 *g* for 5 s to dry the celluloses and then gently transfer the supernatant of **step 8** to the middle of the celluloses of micro-spin column device.
10. Centrifuge the micro-spin column at 10,000× *g* for 5 s.
11. Discard the flow-through and add 500 μL washing solution to the micro-spin celluloses column.
12. Centrifuge the column at 10,000× *g* for 5 s.
13. Repeat **steps 11** and **12** once.
14. Discard the flow-through and centrifuge the column at 10,000× *g* for 30 s to completely eliminate ethanol in cellulose.
15. Place the 0.6 mL centrifuge tube containing the dsRNA-bound celluloses in a new 2.0 mL centrifuge tube, then add 400 μL of 1×STE to the column.
16. Let the column on room temperature for at least 2 min and then centrifuge the column at 10,000× *g* for 5 s.
17. Collect the eluted fraction and discard the 0.6 mL micro-spin cellulose tube (*see* **Note 11**).
18. Add 40 μL of 3 M sodium acetate (pH 5.2) and 1 mL of ethanol to the dsRNA solution (*see* **Note 12**).
19. Incubate at −80 °C for 2 h or overnight and then centrifuge the tube at 20,000× *g* for 5 min to precipitate the dsRNA.

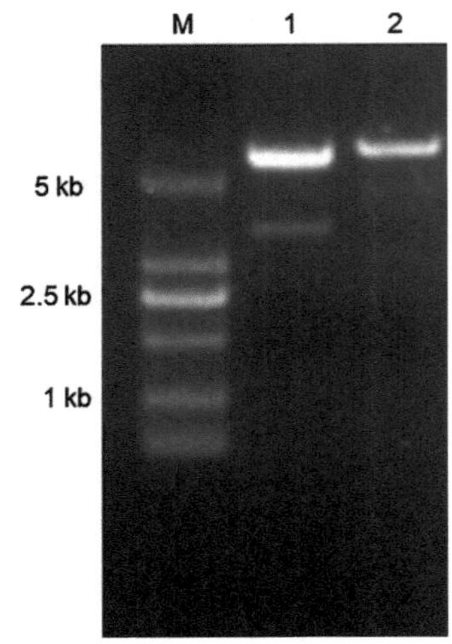

Fig. 2 Gel electrophoresis photo showing the dsRNA profile. dsRNA was extracted from two field samples with mosaic symptoms by the micro-spin cellulose column methods and separated in 1% agarose gel

20. Discard the supernatant and add 500 μL prechilled 75% ethanol.
21. Centrifuge the column at 10,000× *g* for 30 s and discard the supernatant.
22. Repeat **steps 20** and **21** once.
23. Add 500 μL prechilled anhydrous ethanol.
24. Centrifuge the column at 10,000× *g* for 30 s and discard the supernatant.
25. Dissolve the precipitated dsRNA in 20 μL of TE buffer which can be analyzed by agarose gel electrophoresis (Fig. 2).

4 Notes

1. All other solutions should be prepared with DEPC-treated, autoclaved water. DEPC is highly toxic and may dissolve plastic pipettes; therefore, DEPC solution should be measured using a glass sucker in a biosafety fume hood.
2. SDS should be weighed in a biosafety fume hood to avoid inhalation hazard.
3. Do not make too many holes to avoid loss of the cellulose powder during the subsequent high-speed centrifugation.
4. Whatman CF-11 cellulose is no longer available and can be replaced with another cellulose fibrous form with a size between 40 and 100 mesh.

5. The device can be stored at 4 °C for at least one week; however, it is strongly recommended to reprepare each time for best results.
6. Always use symptomatic young leaves and make sure the tissue has been completely homogenized to achieve maximum dsRNA yield.
7. PVPP can be added at about 10–20 mg/mL to the extraction buffer for phenolic compounds-rich plant or fungal tissue.
8. Always use sterile filtered pipette tips.
9. This step can be repeated twice if necessary.
10. DsRNA binds to celluloses typically at about 16–20% ethanol.
11. The micro-spin cellulose tube can be eluted one more time.
12. The following desalting and concentration steps can be omitted if the dsRNA is not used for reverse transcription or library preparation.

References

1. Kalimuthu K, Arivalagan J, Mohan M, Christyraj JRSS, Arockiaraj J, Muthusamy R et al (2022) Point of care diagnosis of plant virus: current trends and prospects. Mol Cell Probes 61:101779. Epub 2021/11/20. https://doi.org/10.1016/j.mcp.2021.101779
2. King AM, Lefkowitz E, Adams MJ, Carstens EB (2011) Virus taxonomy: ninth report of the International Committee on Taxonomy of Viruses. Elsevier
3. Dodds JA, Morris TJ, Jordan RL (1984) Plant viral double-stranded RNA. Annu Rev Phytopathol 22:151–168. Epub September 1984
4. Morris TJ, Dodds JA, Hillman B, Jordan RL, Lommel SA, Tamaki SJ (1983) Viral specific dsRNA: diagnostic value for plant virus disease identification. Plant Mol Biol Rep 1:27–30
5. Valverde R, Nameth S, Jordan R (1990) Analysis of double-stranded RNA for plant virus diagnosis. Plant Dis 74(3):255–258
6. Wu Q, Ding S-W, Zhang Y, Zhu S (2015) Identification of viruses and viroids by next-generation sequencing and homology-dependent and homology-independent algorithms. Annual Rev Phytopathol 53:425–444. Epub 2015/06/07. https://doi.org/10.1146/annurev-phyto-080614-120030
7. Morris TJ, Dodds JA (1979) Isolation and analysis of double-stranded RNA from virus-infected plant and fungal tissue. Phytopathology 69:854–858. Epub 2 March 1979
8. Marais A, Faure C, Bergey B, Candresse T (2018) Viral double-stranded RNAs (dsRNAs) from plants: alternative nucleic acid substrates for high-throughput sequencing. In: Pantaleo V, Chiumenti M (eds) Viral metagenomics: methods and protocols. 1746. New York, Springer, pp 45–53
9. Fukuhara T, Urayama S, Okada R, Kiyota E, Moriyama H (2011) Detection of long and short double-stranded RNAs. In: Kodama H, Komamine A (eds) RNAi and plant gene function analysis: methods and protocols. 744. Humana Press, Totowa, pp 129–144
10. Okada R, Kiyota E, Moriyama H, Fukuhara T, Natsuaki T (2015) A simple and rapid method to purify viral dsRNA from plant and fungal tissue. J Gen Plant Pathol 81(2):103–107. Epub 10 January 2015. https://doi.org/10.1007/s10327-014-0575-6

Chapter 5

Analysis of Virus-Induced Double-Stranded RNA in Living Plant Cells by the dRBFC Assay

Ying Zhang, Xinyue Fan, and Xiaofei Cheng

Abstract

Double-stranded RNA (dsRNA) is the replicate intermediate of all RNA viruses, and is also recognized by their host cells as a virus-invading molecule signal. Analysis of the localization and dynamic of virus-induced dsRNA can reveal crucial information concerning the molecular mechanism of virus replication and host responses to viral infection. In this chapter, we provide an easy and efficient protocol called dsRNA binding-dependent fluorescence complementation (dRBFC) assay for labeling the dsRNAs in living plant cells using two different plant RNA viruses, namely potato virus X and turnip mosaic virus. Moreover, both YFP- and mRFP-based dRBFC plasmids are available for the flexibility of experiment design.

Key words Virus-induced double-stranded RNA, Labeling, dRBFC assay, Living cell

1 Introduction

Double-stranded RNAs (dsRNAs) are double-helical forms of RNA resembling the well-known DNA duplex. dsRNAs can be generated by varied RNA-dependent RNA polymerase (RdRp) using single-stranded RNAs (ssRNAs) as templates, by annealing of two complementary ssRNAs molecules or by self-folding of an internal inverted-repeat sequence within a single RNA [1–4]. DsRNA is usually recognized as the hallmark of viral infection since it is the replication intermediates of all RNA viruses. As a result, nearly all organisms have the capability of recognizing dsRNA and mounting a response. For instance, eukaryotic cells have evolved several dsRNA-associated antiviral mechanisms, e.g., RNA silencing, dsRNA-triggered innate immunity, and dsRNA-specific nucleases [1]. RNA silencing is a conserved across-kingdom sequence-specific mechanism, which is initiated by the cleavage of long dsRNA molecules by type III endonucleases into 21–24 nucleotide short dsRNA duplexes. Subsequently, one strand of the short dsRNA duplexes is then loaded into Argonaute proteins,

Xiaofei Cheng and Guanwei Wu (eds.), *Double-Stranded RNA: Methods and Protocols*, Methods in Molecular Biology, vol. 2771, https://doi.org/10.1007/978-1-0716-3702-9_5,

which recognizes target RNA or DNA through base pairing and lead to transcriptional or posttranscriptional silencing [5]. DsRNA-triggered innate immunity is triggered by the perception of dsRNA by cellular dsRNA receptors such as toll-like receptor 3 (TLR3), MDA-5, RNA helicase RIG-I, and protein kinase PKR, which leads to conventional innate immune responses [6–9]. Additionally, some organisms also encode dsRNA-specific nucleases that target dsRNA for degradation [10, 11]. Understand the subcellular location of dsRNA and its interaction with viral and/or host proteins in living cells can provide considerable information of virus replication and host antiviral responses [8]. In this chapter, we provide a simple and sensitive dsRNA labeling procedure called dsRNA binding-dependent fluorescence complementation (dRBFC) assay in living plant cells.

2 Materials

2.1 Biological Materials

1. *Nicotiana benthamiana* seeds.
2. *Escherichia coli* strains DH5α competent cells.
3. *Agrobacterium tumefaciens* (synonym *Rhizobium radiobacter*) strain GV3101.
4. *Turnip mosaic virus* (TuMV) infectious clone TuMV-6K2mCherry and TuMV::VPg-YFP (*see* **Note 1**).
5. *Potato virus X* (PVX) infectious clone (e.g., pGR107; see **Note 2**).

2.2 Regents and Instruments

1. Pots.
2. Soil mixture (e.g., SunGro Sunshine® Mix #1).
3. Yellow fluorescent protein (YFP)-based dRBFC assay plasmids: pdRBFC-NS1-YC, pdRBFC-NS1-YN, pdRBFC-B2-YN, and pdRBFC-B2-YC (Addgene IDs: 74721-74724) [12] (*see* **Note 3**).
4. Monomer red fluorescent protein (mRFP)-based dRBFC assay plasmids: pdRBFC-B2-NmRFP, pdRBFC-B2-CmRFP, pdRBFC-NS1-NmRFP, and pdRBFC-NS1-CmRFP [13] (*see* **Note 4**).
5. The YFP-tagged PVX TGBp2 construct 35S::YFP-TGBp2.
6. Plasmid purification kit (e.g., Vazyme Biotech FastPure Plasmid Mini Kit; Catalog no. DC201).
7. Regents for Lysogeny broth (LB) medium: Oxoid tryptone (Thermo Scientific, Catalog no. LP0042B), sodium chloride (NaCl; Merck Sigma Aldrich, Catalog no. S9888), Bacto yeast extract (Thermo Scientific, Catalog no. 212750), and Oxoid agar (Thermo Scientific, Catalog no. LP0013T).

8. Antibiotics: kanamycin (Kana; Merck Sigma Aldrich, Catalog no. E004000), Spectinomycin (Spec; Merck Sigma Aldrich, Catalog no. S4014), rifampicin (Rif; Merck Sigma Aldrich, Catalog no. 557303), and gentamycin (Gen; Merck Sigma Aldrich, Catalog no. 345814-M) (*see* **Note 5**).
9. β-(N-Morpholino) ethane sulfonic acid (MES; Merck Sigma Aldrich, Catalog no. M3671).
10. Magnesium chloride ($MgCl_2$; Merck Sigma Aldrich, Catalog no. M1028).
11. Acetosyringone (Merck Sigma Aldrich, Catalog no. D134406).
12. DL15000 DNA ladder (e.g., Vazyme Biotech; Catalog no. MD103-01).
13. Luer-Lok 1 mL syringe (e.g., BD; Catalog no. 309628).
14. Thermocycler (e.g., LongGene T30D (Tri-block Super Gradient) PCR thermal cycler).
15. Shaking incubator (e.g., Yihelab THZ-100).
16. Centrifuge (e.g., Eppendorf 5418 R Centrifuge).
17. Spectrophotometer (e.g., Eppendorf BioPhotometer™ plus and semi-micro Vis Cuvette).
18. Confocal microscopy (e.g., Leica TCS SP8).
19. Electroporation system (e.g., Bio-Rad MicroPulser Electroporator and Electroporation Cuvettes, 0.1 cm gap).
20. Electro-thermal incubator (e.g., Thermo Fisher Scientific Model IMH400-S).
21. Agarose and electrophoresis device (e.g., Liuyi Biotechnology Model 122-3220).
22. Biosafety cabinet (e.g., Airtech Model BSC-03IIA2-SE).
23. Cover glasses (Corning Square and Rectangular Cover Glasses; Catalog no. 12-553-472).

2.3 Medium and Buffers

1. Liquid LB medium: 10 g/L tryptone, 10 g/L sodium chloride (NaCl), 5 g/L yeast extract.
2. Solid LB medium: 10 g/L tryptone, 10 g/L sodium chloride (NaCl), 5 g/L yeast extract, and 1.5% agar.
3. Infiltration buffer: 10 mM $MgCl_2$, 10 mM MES, 150 μM acetosyringone, pH = 5.6.

3 Methods

For impatient people, start from Subheading 3.2.

3.1 Preparation of N. benthamiana Seedlings

1. Grow *N. benthamiana* seedlings in pots in growth chamber or greenhouse with suitable environmental conditions, e.g., 16 h/8 h light/dark photoperiod at 25 °C and 60% humidity.
2. Water plants once a week and fertilize once a week.
3. At 4-week-old stag, stop watering for 2 days before agroinfiltration to allow enough air in leaf spongy parenchyma.

3.2 Preparation of Agrobacteria

1. Transfer TuMV-6K2mCherry, TuMV::VPg-YFP, pGR107, and dRBFC plasmids into agrobacteria via electroporation.
2. Resuspend the agrobacteria with 500 μL liquid LB medium, 20 μL of the agrobacterial mixture was plated onto solid LB plate containing suitable antibiotics (*see* **Note 6**).
3. Incubate the plate at 28 °C until the colony size reaches 1–2 mm.
4. Use polymerase chain reaction (PCR) to select the correct colonies with primers in Table 1 (*see* **Note 7**).
5. Grow correct agrobacteria in liquid LB medium supplemented with appropriate antibiotics under 28 °C at 220 rpm for 1–2 days.
6. Collect bacteria by centrifuging for 1 min at 6000 rpm at room temperature.
7. Resuspend the bacterial pellet in 1 mL infiltration buffer.
8. Repeat **steps 6** and 7 one more time.
9. Measure the O.D. 600 value and adjust the O.D. 600 value to 0.2 and keep them at room temperature for 1–3 h.
10. Infiltrate them into well-expanded *N. benthamiana* leaves using a 1-mL syringe (*see* **Note 8**).

Table 1
Primers used in this protocol

Primer names	Sequence (5′-3′)
TuMV-CP-F	GCAGGTGAAACGCTTGAT
TuMV-CP-R	CAACCCCTGAACGCCCAG
PVX-TGB2-F	ATGTCCGCGCAGGGCCATAGGC
PVX-CP-R	GCGACAGCATTGCTGGCTACTATG
35SF	CTATCCTTCGCAAGACCCTTC
NS1-R	TTTCAGAATCCGCTCCACTATCTG
B2-R	GGCCTTTCCCTCTAGGTATGCCAC
Vp35-R	GATTTTCAGGGCCCGTGTCTCACC

3.3 Confocal Microscope Observation

1. Excise a 0.5 × 0.5 cm leaf patch from the agrobacteria-infiltrated area with a blade at 48–72 h post-infiltration (*see* **Note 9**).
2. Drop 1–2 water drops on the middle of the cover glass (0.13–0.17 mm in thickness).
3. Place the leaf patch on water and cover with another cover glass of the same size.
4. Load cover glass to sample stage and fix it with slide holders.
5. Excite fluorescence with appropriate laser beam (*see* **Note 10**).
6. Collect the emission wavelength between 509 to 550 nm for YFP, and 597 to 645 nm for mRFP or mCherry (*see* **Note 11**).
7. Search cells expressing both YFP and mCherry/mRFP at low magnification (e.g., ×10 objective).
8. Record fluorescence signal at suitable laser intensity, gain, and magnification (*see* **Note 12**).

3.4 Results Analyses

1. In TuMV-6K2mCherry-infected cells, the mCherry fluorescence was recorded on the outer membrane of chloroplasts and irregularly aggregates in the cytosol. The dsRNA fluorescence that was labeled by YFP-based dRBFC assay appears as granules located inside the irregularly 6K2 aggregates, but not on the surface of chloroplasts (Fig. 1a).
2. In TuMV::VPg-YFP-infected cells, YFP can be found in both the nuclei and cytosol. The fluorescence of VPg-YFP in the cytosol appeared as membrane-associated granules, while the dsRNA foci that were labeled by mRFP-based dRBFC assay were located inside the VPg-YFP granules surrounding the nucleus (Fig. 1b).
3. In PVX-infected cells, YFP was found as irregular aggregates surrounding the nuclei and in the cytosol. The dsRNA foci that were labeled by mRFP-based dRBFC assay were wrapped by the YFP-TGBp2 aggregates.

4 Notes

1. The VPg in TuMV::VPg-YFP is C-terminally fused with YFP, which is separated by a short FLAG tag.
2. The pGR107 has been previously described [14].
3. The YFP-based dRBFC plasmids pdRBFC-NS1-YN, pdRBFC-NS1-YC, pdRBFC-B2-YN, and pdRBFC-B2-YC were described previously [12].
4. The mRFP-based dRBFC plasmids have been described [13].

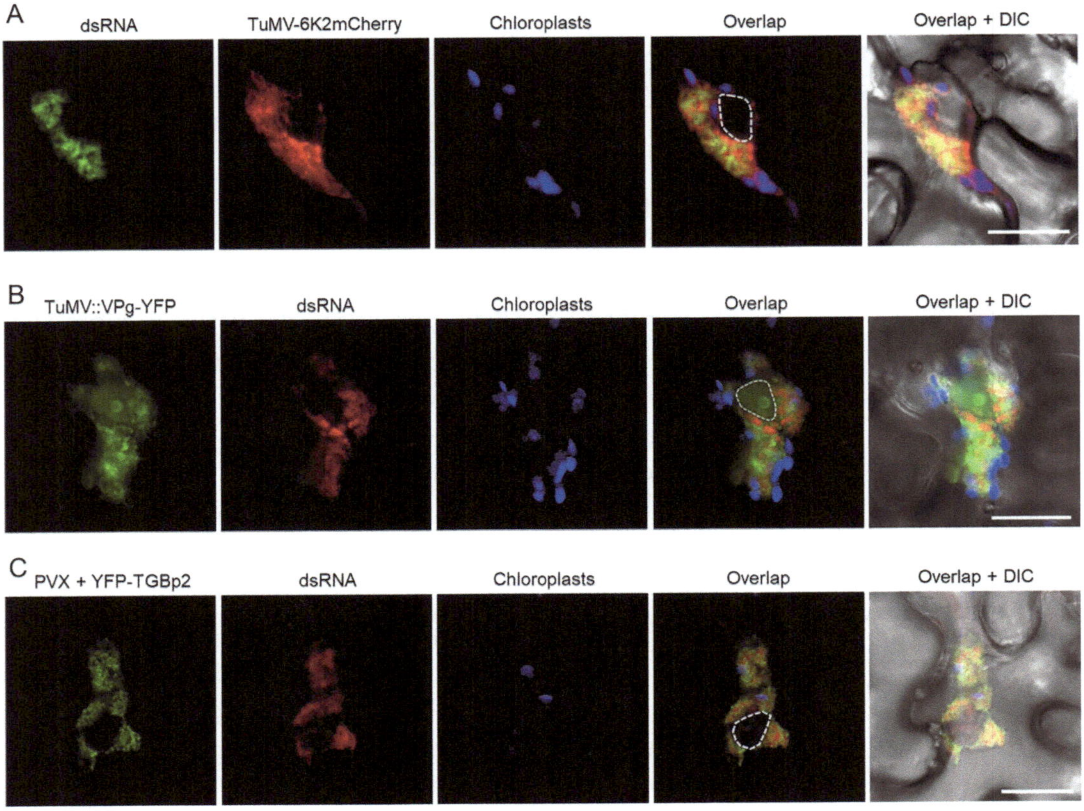

Fig. 1 Visualization of viral dsRNA in *N. benthamiana* epidermal cells by dRBFC assay. Confocal micrographs of *N. benthamiana* epidermal cells infected by TuMV-6K2mCherry (**a**), TuMV::VPg-YFP (**b**), and PVX (**c**) at 60 dpi. The dsRNA was labeled by B2-YN and NS3-YC (**a**) or B2-NmRFP and NS1-CmRFP (**b** and **c**). Dashed area indicates the nucleus. Scale bars = 20 μm

5. Rifampicin is resolved in dimethyl sulfoxide (DMSO) as a 50 mg/mL stock solution and wrapped in aluminum foil, while spectinomycin, kanamycin, and gentamycin are resolved in ddH_2O as 50 mg/mL stock solutions.
6. The mRFP-based dRBFC plasmids have spectinomycin resistance, while TuMV-6K2mCherry, TuMV::VPg-YFP, pGR107, 35S::YFP-TGBp2, and YFP-based dRBFC plasmids have kanamycin resistance. The agrobacteria strain GV3101 has rifampicin and gentamycin resistances.
7. PCR was directly performed with agrobacteria from the plate as the template.
8. When two or more agrobacteria are to be used for co-transformation make sure the OD600 of each construct are identical.
9. Recombinant proteins typically start to express at 36 h post-agroinfiltration.

10. YFP and mRFP/mCherry are usually excited at 488 and 561 nm. But, confocal microscopies of different brands may slightly differ in excitation laser.
11. The emission bandwidth setting cannot include the wavelengths of excitation lasers.
12. Always use minimum laser powder (e.g., 3–5% intensity for YFP and 8–10% intensity for mCherry) and gain setting (e.g., 900 for Leica TCS SP8). Emission bandwidths of CFP, YFP, and RFP are partially overlapped. Therefore, always use the sequential mode when the signal of two or more fluorescent proteins needs to be recorded. Use the minimum gain setting during signal collection to reduce overexposure. Always start with the default and confirm the location of interest is underexposed.

Acknowledgments

This work was supported by grants from the National Natural Science Foundation of China (32022071) to XC.

References

1. Chen YG, Hur S (2022) Cellular origins of dsRNA, their recognition and consequences. Nat Rev Mol Cell Biol 23(4):286–301
2. Sijen T et al (2001) On the role of RNA amplification in dsRNA-triggered gene silencing. Cell 107(4):465–476
3. Rosok O, Sioud M (2004) Systematic identification of sense-antisense transcripts in mammalian cells. Nat Biotechnol 22(1):104–108
4. Dewitte-Orr SJ, Mossman KL (2011) The antiviral effects of extracellular dsRNA. In: Mossman KL (ed) Viruses and interferon: current research. Caister Academic Press, Norfolk, pp 1–18
5. Plasterk RHA (2002) RNA silencing: the genome's immune system. Science 296(5571):1263–1265
6. Peisley A, Hur S (2013) Multi-level regulation of cellular recognition of viral dsRNA. Cell Mol Life Sci 70:1949–1963
7. Jacobs BL, Langland JO (1996) When two strands are better than one: the mediators and modulators of the cellular responses to double-stranded RNA. Virology 219(2):339–349
8. Triantafilou K et al (2012) Visualisation of direct interaction of MDA5 and the dsRNA replicative intermediate form of positive strand RNA viruses. J Cell Sci 125(Pt 20):4761–4769
9. Son K-N, Liang Z, Lipton HL (2015) Double-strand RNA is detected by immunofluorescence analysis in RNA and DNA virus infections including those by negative-strand RNA viruses. J Virol 89
10. Ishibashi K et al (2019) Soybean antiviral immunity conferred by dsRNase targets the viral replication complex. Nat Commun 10(1):4033
11. Cuellar WJ et al (2009) Elimination of antiviral defense by viral RNase III. Proc Natl Acad Sci 106(25):10354–10358
12. Cheng X et al (2015) Visualizing double-stranded RNA distribution and dynamics in living cells by dsRNA binding-dependent fluorescence complementation. Virology 485:439–451
13. Wu X et al (2019) The Potato virus X TGBp2 protein plays dual functional roles in viral replication and movement. J Virol 93(5):e01635–e01618
14. Chapman S, Kavanagh T, Baulcombe D (1992) Potato virus X as a vector for gene expression in plants. Plant J **2**(4):549–557

Chapter 6

Detection of dsRNA with Fluorescence In Situ Hybridization (FISH)

Wenling Zhou, Juan Huang, Xianguang Yang, and Xiaoming Zhang

Abstract

Fluorescence in situ hybridization (FISH) is a powerful technique used for detecting and localizing specific nucleic acid sequences in cells or tissues. Double-stranded RNA (dsRNA) is a type of RNA with complementary strands, highly produced during the replication cycle of RNA viruses. dsRNA plays an essential role in many biological processes, including the activation of RNA silencing. Here, we present an overview of how FISH can be employed to detect and locate dsRNA. The detection and localization of dsRNA through FISH provide valuable insights into RNA-mediated processes and their roles in various biological phenomena.

Key words Fluorescence in situ hybridization, dsRNA detection, Localization

1 Introduction

Fluorescence in situ hybridization (FISH) is a widely utilized technique in molecular biology for the detection of RNA or DNA within cells and tissues [1, 2]. Double-stranded RNA (dsRNA) is a molecule composed of two RNA chains connected by hydrogen bonds, typically found in RNA viruses. Emerging evidence suggests that dsRNA is not limited to virus-infected cells, but also exists in aberrant immune responses. The discovery and characterization of dsRNA have revolutionized our understanding of RNA-mediated processes. FISH enables the visualization and determination of the spatial distribution of dsRNA molecules within cells and tissues [3]. FISH has become an invaluable tool for investigating the location and abundance of dsRNA. It plays a pivotal role in enhancing our understanding of the functions of dsRNA in various biological processes, including gene regulation, RNA interference, and viral replication. The development and application of FISH in dsRNA detection hold great promise for advancing our knowledge of RNA biology and its implications in health and disease [4].

Xiaofei Cheng and Guanwei Wu (eds.), *Double-Stranded RNA: Methods and Protocols*, Methods in Molecular Biology, vol. 2771,
https://doi.org/10.1007/978-1-0716-3702-9_6,

Here, we describe the details of dsRNA for FISH detection. We adapt the system for detecting dsRNA by direct fluorescence labeling of fixed samples.

2 Materials

1. The *Agrobacterium tumefaciens* strain GV3101 (SGn306) is purchased from Shanghai Shifeng Biotechnology Co.
2. The primary antibody for dsRNA (J2) is purchased from Scicons.
3. The Alexa Fluor 488 goat anti-mouse IgG antibody (62197) is purchased from Sigma-Aldrich.
4. The poly-L-lysine pre-coated coverslips (P4832) are purchased from Sigma-Aldrich.
5. LB medium: 25 μg/mL rifampicin, 50 μg/mL gentamicin, and 25 μg/mL kanamycin.
6. Infiltration buffer: 10 mM $MgCl_2$, 0.15 mM Acetosyringone, and 10 mM 2-Morpholinoethanesulfonic Acid (MES).
7. Enzyme solution: 0.4 M mannitol, 20 mM KCl, 20 mM MES, 10 mM $CaCl_2$, 1.5% cellulase R10, 0.4% macerozyme R10, and 0.1% BSA.
8. W5 solution: 2 mM MES (pH 5.7), 5 mM KCl, 154 mM NaCl, and 125 mM $CaCl_2$.
9. MMG solution: 0.4 M mannitol, 15 mM $MgCl_2$, and 4 mM MES.
10. PEG solution: 40% PEG4000, 100 mM $CaCl_2$, and 200 mM mannitol.
11. M&Ms buffer: 550 mM Mannitol, 5 mM $MgCl_2$, and 2 mM MES.
12. 1×PBST: PBS supplemented with 0.1% Tween-20.
13. 0.1% sodium borohydride solution: ddH_2O supplemented with 0.1 g sodium borohydride.

3 Methods

3.1 Viral Infection

1. The *A. tumefaciens* strain GV3101 carrying the genome of *TuMV* are cultured at 28 °C in LB medium.
2. The *A. tumefaciens* GV3101 carrying TuMV constructs are resuspended with infiltration buffer to a final concentration of OD600 = 0.5.

3. For dsRNA detection, the plant leaves (4-week-old *Nicotiana benthamiana*) are infected with *TuMV* for 48 h, and then collected and isolated protoplast (*see* **Note 1**).

3.2 Protoplast Preparation

1. The equivalent parts of plants are cut into 0.5–1 mm strips with fresh razor blades (*see* **Note 2**).
2. Then, the leaf strips are removed quickly and gently into the enzyme solution, and incubated in the dark at 28 °C (40 rpm) for 4 h.
3. Protoplasts are filtered through 40 mum nylon meshes with 3 washes using an equal volume of W5 solution.
4. After centrifugation at 100 g at 4 °C for 2 min, the protoplasts are suspended in a pre-cooled W5 solution and incubated on ice for 30 min.

3.3 Fixation

1. Protoplasts are washed and resuspended in M&Ms buffer.
2. Before adding 0.05% Triton-X100 for 10 min, protoplasts are fixed in 4% paraformaldehyde (PFA) final concentrations in 4 mM MES and 10 mM EGTA for 20 min.
3. Protoplasts are then washed once with M&Ms buffer, and three times in increasing PBS concentrations from one-third to 1×.
4. Protoplasts are fixed on the poly-L-lysine pre-coated coverslips, and treated for 1 h with 0.1% sodium borohydride solution (*see* **Note 3**).

3.4 Hybridization

1. The samples are incubated with PBST containing 2% BSA for 4 h at 4 °C to block unspecific binding sites.
2. The samples are incubated with primary antibody (such as J2) for 1 h at room temperature (*see* **Note 4**). The antibody is used at 1:200 dilution in 2% BSA to detect dsRNA.
3. After washing with PBS three times for 5 min, the samples are incubated with a mixture of Alexa Fluor 488 goat anti-mouse IgG antibody (1:300) in the dark for 2 h at room temperature (*see* **Note 5**).
4. The samples are washed three times with PBST for 5 min each time, twice with PBS for 5 min each time, and once with distilled H_2O for 1 min.

3.5 Observation

The results are observed using a fluorescence microscope under visible light illumination (Fig. 1).

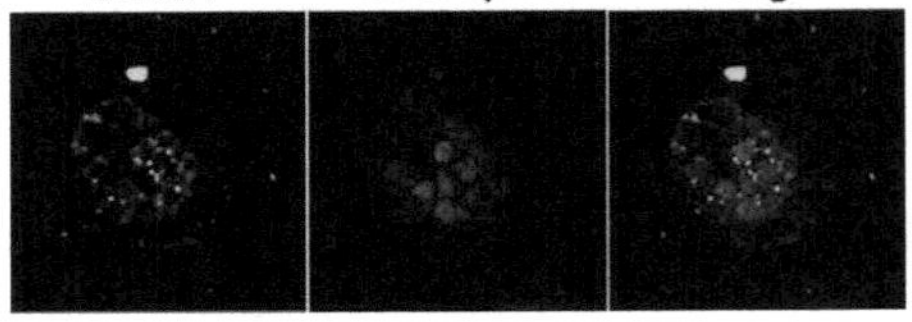

Fig. 1 Specific detection of viral dsRNA-species by FISH. The green dots represent dsRNA, and the red fluorescent signal indicates chloroplasts

4 Notes

1. Setting up a control group is indeed necessary when conducting virus infection experiments or any experimental study in general.
2. Samples preparation for protoplast are often applied to young and tender plant tissues.
3. During the fixation of the protoplasts, the action is as gentle as possible, avoiding protoplasts damage.
4. To ensure binding quality and prevent dryness, antibody incubation should be carried out in a wet box as much as possible.
5. Light should be avoided during the incubation of secondary antibody to prevent fluorescence quenching.

Acknowledgments

This work is supported by National Natural Science Foundation of China (NSFC 32090012 and 32325042), Program of CAS (ZDBS-LY-SM027), and Hainan Yazhou Bay Seed Lab (B21HJ0104, XZ).

References

1. Ramos JM (2022) Fluorescent In Situ Hybridization (FISH). Methods Mol Biol 2422:179–189. https://doi.org/10.1007/978-1-0716-1948-3_12
2. Vyboh K, Ajamian L, Mouland AJ (2012) Detection of viral RNA by fluorescence in situ hybridization (FISH). J Vis Exp 63:e4002. https://doi.org/10.3791/4002
3. Monsion B, Incarbone M, Hleibieh K, Poignavent V, Ghannam A, Dunoyer P, Daeffler L, Tilsner J, Ritzenthaler C (2018) Efficient detection of long dsRNA in vitro and in vivo using the dsRNA binding domain from FHV B2 protein. Front Plant Sci 9:70. https://doi.org/10.3389/fpls.2018.00070
4. Cassidy A, Jones J (2014) Developments in in situ hybridisation. Methods 70(1):39–45. https://doi.org/10.1016/j.ymeth.2014.04.006

Chapter 7

Subcellular Colocalization Assay of Host Factors with Viral Replication Complex in the dsRNA Reporter *Nicotiana benthamiana*

Xinxin Fang, Jianping Chen, Fei Yan, and Guanwei Wu

Abstract

As obligate pathogens, plant viruses co-opt several host factors for viral replication. Double-stranded RNA (dsRNA) plays important roles in plants, including eliciting innate immune responses and RNA interference. dsRNA also represents the genetic entities of a number of viruses and is a marker of infection by positive-sense single-stranded RNA viruses. Previous detection methods for RNA viruses basically relied on immunological methods, but later researchers discovered that the dsRNA-binding domain of the Flock house virus B2 protein is a perfect alternative to the J2 mAb for sensitive and rapid detection of long dsRNA in vitro and in vivo, and developed B2:GFP transgenic *Nicotiana benthamiana* line. This method describes in detail how to visualize host factors in the viral replication complex in time and space with the help of B2: GFP transgenic plants, exemplified by *Turnip mosaic virus* (TuMV), a representative virus member of the Potyviruses.

Key words Confocal, *Turnip mosaic virus* (TuMV), B2:GFP, *Nicotiana benthamiana*, Host factors, Viral replication complex (VRC)

1 Introduction

Double-stranded RNA (dsRNA), which has two complementary strands of RNA, is considered to be the earliest form of life [1, 2]. And usually, this type of RNA is found in the genomes of double-stranded RNA viruses, and is therefore also considered a universal marker of viral infection of cellular organisms [3, 4].

With viruses damaging plant biology, plants have developed a complex and meticulous set of defense mechanisms, such as innate immunity and RNA interference [5, 6], in the course of evolution to resist viral infestation. Based on the fact that viral-derived dsRNA harms plants and causes disorders in various systems within the plant, an effective dsRNA detection tool would help researchers

Xiaofei Cheng and Guanwei Wu (eds.), *Double-Stranded RNA: Methods and Protocols*, Methods in Molecular Biology, vol. 2771, https://doi.org/10.1007/978-1-0716-3702-9_7,

to further understand the roles of dsRNA. Most of the current detection techniques are based on immunological techniques, using antibodies as probes to recognize dsRNA [3, 7–9]. For a long time, J2 mAb became the gold standard for detecting dsRNA [10]; later, 9D5 mAb and FHV B2 mAb gradually shook the dominant position of J2 mAb due to their better binding ability [9]. Although these immunological approaches are effective in recognizing dsRNA, they also suffer from the great drawback of not being able to achieve spatiotemporal visualization of dsRNA in vivo. In contrast, the emergence of B2: GFP transgenic plant has made this limitation almost disappear [11, 12].

As one of the most studied viruses in the Potyviruses, the 53-amino acid membrane protein 6K2 in TuMV can target and induce endoplasmic reticulum proliferation, forming a viral replication complex (VRC) at the exit site of the endoplasmic reticulum [13, 14]. Here, we present a detailed protocol showing the subcellular colocalization assay of the fluorescently tagged host factors with the viral replication complex in *N. benthamiana* plants stably expressing B2:GFP. This system will be a powerful tool to identify host factors colocalizing or associated with VRC.

2 Materials

2.1 Plasmids

pCB301 containing the TuMV infectious clone under the control of 35S promoter was constructed and kept in our lab. pCV-RFP vector was used for RFP-tagged host gene expression vector construction.

2.2 Plant Material and Bacterial Strains

1. 3- to 4-week-old transgenic *N. benthamiana* plants B2:GFP plants were grown in greenhouse at 22 °C with a light cycle of 16 h, a dark cycle of 8 h, and a relative humidity of 60% (*see* **Note 1**).
2. *Escherichia coli (E. coli)* DH5α strain was used for molecular cloning and plasmid amplification, and *Agrobacterium tumefaciens (A. tumefaciens)* GV3101 for electroporation transformation and agroinfiltration.

2.3 Equipment

1. Thermostatic incubator.
2. Autoclave machine.
3. 50 mL Falcon tubes.
4. pH meter.
5. 0.45 μm syringe filters.
6. Petri dishes.

7. Pipettes and tips.
8. 1.5 mL and 2.0 mL Eppendorf Tubes.
9. Eppendorf Centrifuge 5420.
10. Vortex Oscillator.
11. Gene Pulser/MicroPulser Electroporation Cuvettes, 0.2 cm gap.
12. Eppendorf BioPhotometer.
13. Electrophoresis equipment.
14. Nikon A1R HD25 confocal microscope (Nikon Microsystems, Japan).
15. Slides and coverslips.
16. Plant Leaf Sampler and Tweezers.

2.4 Reagents

1. LB liquid medium: Weigh 5 g yeast extract, 10 g peptone, 10 g NaCl into 800 mL distilled water and adjust the pH to 7.0 with NaOH, then volume to 1 L with distilled water and sterilize at 121 °C for 20 min in an autoclave.
2. LB solid medium: 1 L LB liquid medium was mixed with 15 g agar and autoclaved.
3. 50×TAE (Tris-Acetate-EDTA) buffer: 242 g Tris-base was dissolved in 700 mL distilled water, 57.1 mL glacial acetic acid and 100 mL 0.5 M EDTA solution (pH = 8.0, 18.61 g EDTA sodium salt was dissolved in 100 mL of distilled water, and the pH was adjusted with NaOH) was added and then it was concentrated to 1 L. The concentration of the working solution was 1×.
4. 1 M MES (2-(*N*-morpholino) ethanesulfonic acid)): weigh 19.524 g MES dissolved in 100mL distilled water, pH adjusted to 5.6 with NaOH, filtered and sterilized, stored at 4°C.
5. 1 M $MgCl_2$ solution: weigh 20.33 g $MgCl_2 \cdot 6H_2O$ dissolved in 100 mL distilled water. Filter sterilized and stored at 4 °C for spare parts.
6. 200 mM As (Acetosyringone) solution: dissolve 1.0 g of Acetosyringone in 25.48 mL of DMSO, mix thoroughly, and then dispense 1.8 mL of each solution and store at −40 °C.
7. Agrobacterium infiltration buffer (500 mL system): dissolve 5 mL of 1 M $MgCl_2$, 5 mL of 1 M MES, 500 μL of 200 mM As in 500 mL of sterile water, mix well and store at 4 °C (*see* **Note 2**).

3 Methods

3.1 Construction of the Red Fluorescent Protein-Tagged Host Gene via Ligation-Independent Cloning Method

1. Primer design: Design specific primers according to the corresponding CDS of the host gene.
2. The host gene cloning: High-fidelity enzymes of PrimeSTAR series from TAKARA were selected for PCR. 50 μL reaction system consisted of 25 μL of PrimeSTAR® GXL Premix (2×), 1.5 μL of specific upstream primer F, 1.5 μL of specific downstream primer R, 2.0 μL of template cDNA, and 20 μL of ddH_2O. The reaction conditions were 98 °C, 2 min pre-denaturation; (98 °C, 15 s; 55 °C, 20 s; 68 °C, 2 kb/min)/cycle; 68 °C, 8 min final extension, and generally 30~35 cycles were performed.
3. Gel electrophoresis: the product after PCR was added with 6 μL of 5×DNA loading buffer and mixed well, and then run in 1% agarose gel electrophoresis at 200 V for 15 min; then compare whether the size of the bands is correct or not according to the Marker under UV.
4. Gel Recovery: Gels with the correct band size were cut off with a scalpel blade under UV and subjected to gel recovery. PCR products containing the target gene were purified and recovered using OMEGA's Gel Extraction Kit.
5. The resulting target product with A treatment: 5 μL of recovered product, 1 μL of 10×buffer, 0.5 μL of 100 nM dATP, 0.2 μL of T4 DNA Polymerase, and 3.3 μL of ddH_2O were mixed well and subjected to the reaction at 37 °C, 20 min; 75 °C, 20 min in PCR instrument.
6. Processing of pCV-RFP vector: The prokaryotic expression vector pCV-RFP plasmid available in the laboratory was enzymatically digested with the Fast Digest ApaI from Thermo-Fisher, and the reaction system and time were performed according to the instructions.
7. Fragment recovery of the digested vector was also performed with the Gel Extraction Kit from OMEGA.
8. Add T treatment for the recovered vector: 50 μL of recovered product, 10 μL of 10×buffer, 5 μL of 100 nM dTTP, 2 μL of T4 DNA Polymerase, and 33 μL of ddH_2O were mixed well and then subjected to the reaction at 37 °C for 20 min; 75 °C for 20 min in PCR machine.
9. The target gene was ligated to the vector: Take 2 μL each of the host gene plus A and the vector plus T, mix them and perform the reaction in the PCR instrument, the reaction process was as follows: 70 °C, 20 min; 22 °C, 20 min.

10. Transformation: All ligation products were transferred into *E. coli* receptor cells (DH5α), coated on kanamycin-resistant LB solid plates, and cultured overnight at 37 °C in constant temperature with shaker shocks.
11. Identification of positive clones: Pick single colonies for PCR verification and select positive bacteria for sequencing. Positive clones are defined as those with correct sequencing results.
12. Plasmid extraction: Positive mono-colonies were expanded and cultivated, and then plasmids were extracted according to the instructions of OMEGA Plasmid Mini Kit I to obtain RFP-tagged host gene plasmids.

3.2 Introduction of Viruses and Host Genes into Plant Epidermal Cells Through Agroinfiltration

1. pCB301-TuMV infectious clone or RFP-tagged host gene plasmids were transfected into *Agrobacterium tumefaciens* GV3101 by electroshocking method and inoculated onto LB agar plates containing corresponding antibiotics, then the plates were inverted and incubated in a constant temperature incubator at 28 °C for 48 h.
2. Single colonies were picked for PCR verification to select positive clones, which were then added to LB liquid medium containing corresponding antibiotics and incubated overnight at 200 rpm in a constant temperature incubator at 28 °C.
3. Centrifuge the cells at 5000 rpm for 5 min at room temperature to collect Agrobacterium cells.
4. At the end of centrifugation, decant the supernatant and resuspend the precipitated cells with Agrobacterium Infiltration Buffer.
5. Repeat **steps 3** and **4** twice.
6. Adjust the final OD_{600} of Agrobacterium to 0.2 for pCB301-TuMV and RFP-tagged host genes with Agrobacterium infiltration buffer.
7. Infiltration of both Agrobacterium cells harboring TuMV and RFP-tagged host factors or RFP-tagged host genes alone into B2:GFP transgenic plants using a 1 mL syringe.

3.3 Confocal Visualization of RFP-tagged Host Factor Localization with VRC in N. benthamiana B2:GFP

1. At 3 dpi, proceed confocal analysis of RFP-X localization with B2:GFP-labelled VRC. Usually no cell-to-cell movement occurs at this time point for TuMV [15] (*see* **Notes 3** and **4**).
2. At 3 dpi, in B2:GFP leaf tissues infiltrated with RFP-X, the green fluorescence is only observed in the nucleus (*see* **Note 5**).
3. At 3 dpi, in B2:GFP leaf tissues pre-infiltrated with pCB301-TuMV, a few cells may be infected by TuMV. As TuMV replications at ER [14], thus, we can observe apparent punctate bodies emitted by B2:GFP around the nucleus, indicating that this cell has been infected by TuMV and B2 binds with VRC.

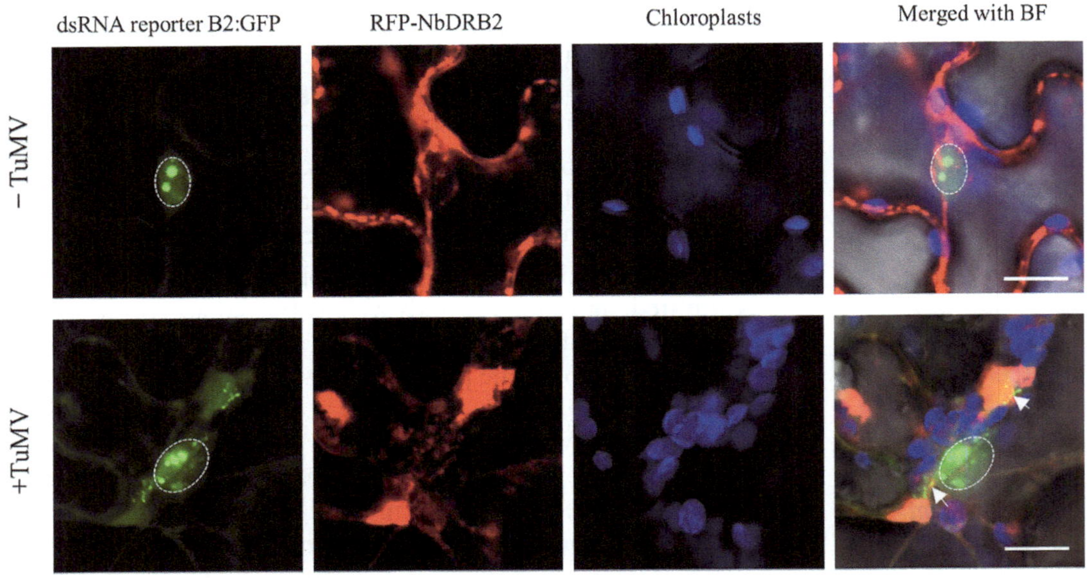

Fig. 1 Visualization of the dsRNA replication intermediates of TuMV. Colocalization assay of the host protein NbDRB2 fused with RFP tag in B2:GFP plants in the presence of TuMV infection (lower panels) or not (upper panels). Representative colocalization signals are indicated by white arrows. Nucleus are indicated with white ovals. Scale bar, 10 μm. BF, bright field

4. As shown in Fig. 1, we can clearly observe that dsRNA signals resulting from VRCs clearly colocalized with RFP-tagged dsRNA-binding protein 2 from *N. benthamiana* (NbDRB2) under TuMV infection. In the absence of viral infection, the RFP-tagged NbDRB2 did not co-localize with the dsRNA signals produced by B2:GFP, which mainly accumulates in the nucleus. This result suggests that NbDRB2 is recruited into the TuMV VRC.

4 Notes

1. Maintaining good watering habits to avoid drought stress of water deprivation in plants before inoculation.
2. The pH of the Agrobacterium infiltration buffer is important for Agrobacterium infection, so it is important to adjust the pH.
3. TuMV consists of a relatively large genome, thus, requiring a longer time to accumulate enough dsRNA for B2:GFP labelling. Thus, we recommend inoculating TuMV one day in advance prior to RFP-X agro-infiltration at the same position.
4. Use the sequential scanning mode to avoid/reduce the cross-talk when multiple fluorophores are analyzed.

5. B2:GFP was found to be highly enriched in the nucleolus and in nucleoplasmic speckles, which likely represent nuclear dicing bodies enriched in DCL1 and DRB4. As most RNA viruses replicate in cytoplasm, thus, these signals do not affect further colocalization analysis of host factors with VRC.

Acknowledgments

We are indebted to Prof. Christophe Ritzenthaler (Universite de Strasbourg, France) for the transgenic T2 dsRNA reporter *N. benthamiana* line B2-GFP. This research was funded by grants from the National Natural Science Foundation of China (3207016) to GW, the China Agriculture Research System of MOF and MARA (CARS-24-C-04) to FY, and the K. C. Wong Magna Fund in Ningbo University.

References

1. Gilbert W (1986) Origin of life: the RNA world. Nature 319:618–618
2. Joyce GF (1989) RNA evolution and the origins of life. Nature 338:217–224
3. Morris TJ, Dodds JA (1979) Isolation and analysis of double-stranded RNA from virus-infected plant and fungal tissue. Phytopathology 8:854–858
4. Kawai T, Akira S (2006) Innate immune recognition of viral infection. Nature Immunology 7:131–137
5. Takeuchi O, Akira S (2009) Innate immunity to virus infection. Immunol Rev 227:75–86
6. Goubau D (2013) Cytosolic sensing of viruses. Immunity 38:855–869
7. Poynter SJ, DeWitte-Orr SJ (2017) Visualizing virus-derived dsRNA using antibody-independent and -dependent methods. Methods Mol Biol 1656:103–118
8. O'Brien CA (2015) Viral RNA intermediates as targets for detection and discovery of novel and emerging mosquito-borne viruses 9: e0003629
9. Son KNZ (2015) Double-stranded RNA is detected by immunofluorescence analysis in RNA and DNA virus infections, including those by negative-stranded RNA viruses. J Virol 89:9383–9392
10. Schönborn J (1991) Monoclonal antibodies to double-stranded RNA as probes of RNA structure in crude nucleic acid extracts. Nucleic Acids Res 19:2993–3000
11. Chao JA (2005) Dual modes of RNA-silencing suppression by Flock House virus protein B2. Nature Structural & Molecular Biology 12: 952–957
12. Monsion B (2018) Efficient detection of long dsRNA in vitro and in vivo using the dsRNA binding domain from FHV B2 protein. Front Plant Sci 9:70–70
13. Laliberté JF, Sanfaçon H (2010) Cellular remodeling during plant virus infection. Annu Rev Phytopathol 48:69–91
14. Wei T, Wang A (2008) Biogenesis of cytoplasmic membranous vesicles for plant potyvirus replication occurs at endoplasmic reticulum exit sites in a COPI- and COPII-dependent manner. J Virol 82:12252–12264
15. Cui XY (2017) The C-terminal region of the Turnip mosaic virus P3 protein is essential for viral infection via targeting P3 to the viral replication complex. Virology 510:147–155

Chapter 8

Production of Double-Stranded RNA Using the Prokaryotic Promoter-Mediated Bidirectional Transcription

Xue Jiang, Kekely Bruno Attiogbe, Yating Guo, and Xiaoyun Wu

Abstract

Large-scale and cost-less production of double-stranded RNA (dsRNA) is the basis for the widespread application of dsRNA in agriculture. Bidirectional transcription of target sequence in RNase III-deficient *Escherichia coli* strain HT115 (DE3) is an efficient way to produce large amounts of dsRNA. Here, we present a detailed method for the production of dsRNA by bidirectional transcription in *E. coli* from vector construction, induction of expression by isopropylthio-β-galactoside (IPTG), and purification of dsRNA from *E. coli*.

Key words Prokaryotic promoter, dsRNA, Bidirectional transcription, T7, RNase III-deficient

1 Introduction

RNA silencing (also known as RNA interference, RNAi) is an evolutionarily conserved sequence-specific gene-inactivation system and antiviral mechanism [1]. Double-stranded RNA is the trigger of RNA silencing, which is first recognized by Dicer in animals [or Dicer-like (DCL) in plants], a dsRNA-specific endonuclease of the RNase III family, and cleaved into 21-24 nucleotide (nt) RNAs, termed as small interfering RNAs (siRNAs) or microRNAs. These small RNAs were then cooperated into Argonaute (AGO) proteins, which further form multiple protein complexes termed RNA-induced silencing complex (RISC) for function through base pairing between small RNA and target DNA or RNA [2]. Nowadays, RNAi has become a powerful approach to suppress or silence the expression of target genes in plant, insect, or fungi for practice application or gene functional study. Several methods have been developed to trigger RNAi in plants, e.g., expressing artificial microRNA gene [3], virus-induced gene silencing [4], and transgenic expression of hairpin structure RNA transcript [5]. Recently, several studies showed that directly spray of

Xiaofei Cheng and Guanwei Wu (eds.), *Double-Stranded RNA: Methods and Protocols*, Methods in Molecular Biology, vol. 2771, https://doi.org/10.1007/978-1-0716-3702-9_8,

dsRNA also can induce RNAi in plants, fungi, and insects [6–9]. However, the application of dsRNA as a pesticide or fungicide is hampered by the question of how to acquire dsRNA in large quantities and cheaply.

Due to easy and costless in propagation, clear genetic background, and convenient genetic manipulation, *Escherichia coli* has become one of the most important model organisms in molecular biology. *E. coli* was widely used for protein expression [10, 11]. Recently, dsRNA was also successfully produced via bidirectional transcription in the RNase III-deficient *E. coli* strain HT115 (DE3) [12]. Bidirectional transcription contains steps including the construction of an expression vector containing the target gene, transformation of the vector into a RNase III-deficient *E. coli* strain, induction of the transcription by IPTG, and finally purification of dsRNA [13, 14]. In this chapter, we provide a detailed procedure for the production of dsRNA in *E. coli* with the expression vector pET17b under the T7 promoter (Fig. 1). The subsequent dsRNA purification procedure is also included for convenience.

2 Materials

2.1 Primer Design and Amplification of Target Fragment

1. The empty pET-28a plasmid. Store at −20 °C (*see* **Note 1**).
2. DNA polymerase (e.g., 2×Phanta Max Master Mix dye plus; Cat no. P525-01). Store at −20 °C.
3. 50×TAE buffer: Weight 242 g Tris base, 37.2 g EDTA disodium salt dihydrate, 57.1 mL Acetic acid, top up with autoclaved, deionized water to 1 L and store at room temperature.
4. Gel DNA Extraction kit (e.g., Vazyme FastPure Gel DNA Extraction Mini Kit; Cat no. DC301-01).
5. Absolute ethanol.
6. Thermal cycler (e.g., LongGene T30 Tri-block Gradient Thermal Cycler).
7. Pipette.
8. 2×RNA Loading Buffer (e.g., New England Biolabs RNA loading dye, Cat. no. B0363S).
9. Desktop centrifuge.
10. Autoclaved, deionized water.
11. Agarose.
12. Gel electrophoresis device.
13. Ultraviolet illuminator.
14. Microwave oven.
15. Flask.
16. Agar.

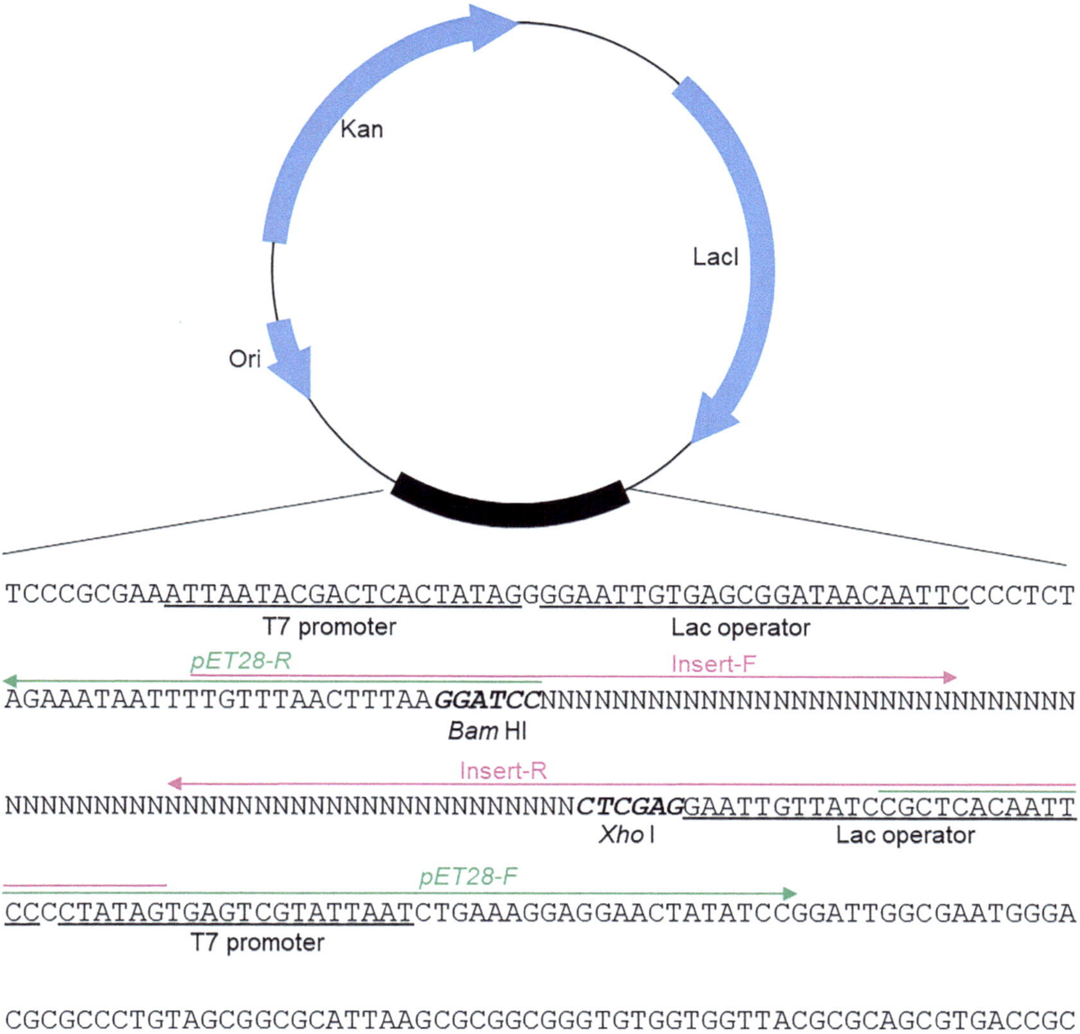

Fig. 1 Schematic diagram of the modified pET28-a vector for bidirectional transcription. The T7 promoter, lac operator, and restriction sites of *Bam* HI and *Xho*I are shown underlined or in italic. The primers used to amplify vector and insertion are also shown as green and pink arrows, respectively

2.2 Clone and Transformation

1. Recombinant cloning kit (e.g., Vazyme ClonExpress II One Step Cloning Kit; Cat no. C112-01).
2. Competent cells of *Escherichia coli* DH5α. Store at −80 °C (*see* **Note 2**).
3. LB medium:10 g/L Tryptone,10 g/L Yeast extract, 5 g/L NaCl, solid medium plus 15 g/L agar powder, sterilize at 121 °C for 20 min (*see* **Note 3**).
4. Kanamycin.
5. Water bath.
6. Lab incubator (e.g., Thermo Scientific BK6160).
7. Incu-Shaker (e.g., Benchmark Scientific Incu-Shaker Mini).

2.3 Expression and Isolation of dsRNA

1. Competent cells of RNase III-deficient *Escherichia coli* strain HT115 (DE3). Store at −80 °C.
2. 2×YT medium: 16 g/L Tryptone, 10 g/L Yeast extract, 5 g/L NaCl, sterilize at 121 °C for 20 min.
3. 1 M isopropyl β-D-1thiogalactopyranoside (IPTG): Weigh out 2.38 g IPTG, add deionized water to the volume 10 mL, filter sterilization with a 0.22 μm filter, aliquot and store at −20 °C.
4. 1 M Ammonium acetate (NH_4OAc).
5. 0.5 M EDTA (pH 8.0) solution: Weigh 1.861 g $Na_2EDTA{\cdot}2H_2O$, add about 8 mL of deionized water and stir well, adjust the pH to 8.0 with NaOH, add deionized water to set the solution to 10 mL (*see* **Note 4**).
6. Phenol/Chloroform/Isoamyl alcohol: After mixing Tris-HCl (pH 5.6) balanced phenol with an equal volume of chloroform/isoamyl alcohol (24:1), transfer to a brown glass bottle and store at 4 °C.
7. Anhydrous ethanol.
8. DNase I (e.g., Vazyme DNase I, RNase-free; Cat no. EN401). Store at −20 °C.
9. RNase A (e.g., Vazyme RNase A; Cat no. DE111).

3 Methods

All procedures are carried out at room temperature unless emphasized otherwise.

3.1 Clone of Target Genes

1. Design the primers to linearize the pET-28a vector and amplify target DNA fragment (*see* **Note 5**).
2. Perform the polymerase chain reaction (PCR) to amplify the pET-28a vector and target DNA fragment (*see* **Note 6**).
3. Analyze the PCR product by electrophoresis in a 1% agarose gel.
4. Under the UV illuminator, excise the agarose gel containing target amplicon with a clean scalpel and transfer to a 1.5-mL centrifuge tube.
5. Add 500 μL Buffer GDP (Provided in the FastPure Gel DNA Extraction Mini Kit) to completely cover the gel strip (*see* **Note 7**).
6. Heat the centrifuge tube containing the gel strip in a metal bath at 55 °C for 10–20 min, during which gently upside down several times to promote dissolution.
7. Transfer the gel solution to the spin column with a 2.0-mL collection tube.

8. Centrifuge at 12,000 rpm for 1 min, discard the flow-through and place the spin column back to the same 2.0-mL collection tube.
9. Add 600 μL Buffer GW (Provided in the FastPure Gel DNA Extraction Mini Kit) to the spin column (*see* **Note 8**).
10. Centrifuge at 12,000 rpm for 1 min, and discard the flow-through and place the spin column back to the same 2.0-mL collection tube.
11. Repeat **steps 9** and **10** once.
12. Centrifuge at 12,000 rpm for 1 min to completely eliminate residual wash buffer in the spin column.
13. Transfer the spin column to a new, clear 1.5-mL centrifuge tube, and add 20 μL of sterilized water.
14. Incubate at 55 °C for 2 min, and centrifuge at 12,000 rpm for 1 min.
15. Discard the spin column, and the elusion product can be used for recombinant cloning immediately or stored at −20 °C until use (*see* **Note 9**).

3.2 Vector Construction

1. Set up the recombinant cloning reaction mix: Add 2 μL 5×CEII Buffer, 1 μL Exnase II, add deionized water to 10 μL (*see* **Note 10**).
2. Incubate the mixture at 37 °C for 30 min; meantime, thaw competent cells of *E. coli* strain DH5α on ice.
3. Add the above ligation product into the *E. coli* competent cells, gently mix well and incubate on ice for 20 min.
4. Incubate the tube containing *E. coli* competent cells at 42 °C for 45 s and place back on ice immediately for 2 min (*see* **Note 11**).
5. Add 500 μL liquid LB to the tube containing *E. coli* competent cells, incubate at 37°C in a horizontal shaker for 1 h.
6. Centrifuge at 6000 rpm for 30 s to collect the bacteria.
7. Discard most supernatant and resuspend the bacteria pellet by pipetting (*see* **Note 12**).
8. Plate half of the transformation onto a 10 cm LB agar plate containing 50 μg/mL kanamycin.
9. Incubate the plate overnight at 37 °C.
10. Select the correct colonies by PCR.
11. Grow the correct colonies in liquid LB medium at 37 °C and 120 rpm overnight.
12. Purify the plasmid with commercial plasmid extraction kit (*see* **Note 13**).
13. Confirm the plasmid by Sanger sequencing.

3.3 Expression of dsRNA

1. Transform the recombinant vector into *E. coli* strain HT115 (DE3) and select the correct colony as described in Subheading 3.2.
2. Culture the correct colony in 4 mL LB medium supplied with 50 μg/mL kanamycin at 190 rpm in a Incu-Shaker at 37 °C overnight.
3. Take 1 mL of the overnight bacterial culture into 50 mL fresh YT medium containing 50 μg/mL kanamycin.
4. Incubate the medium at 37 °C and 190 rpm for about 4 h until the O.D.600 reached about 0.4 (*see* **Note 14**).
5. Add IPTG to the final concentration of 1.0 mM to the culture (*see* **Note 15**).
6. Induct the transcription for additional 5 h under the same conditions (*see* **Note 16**).
7. Transfer the culture to a 50 mL centrifuge tube, centrifuge at 5000× *g* for 10 min at 4 °C to collect the bacteria.
8. Discard the supernatant after centrifugation and the bacterial pellet can be used for dsRNA extraction immediately or temporarily stored at −80 °C until use.

3.4 Isolation of dsRNA

1. Measure 2 mL of 1 M NH_4OAc and 40 μL 0.5 M EDTA (pH 8.0), mix well and resuspend the bacteria pellet in the mixture using a 1-mL pipette (*see* **Note 17**).
2. Add an equal volume of phenol/chloroform/isoamyl-alcohol mixture (25:24:1, v/v) to the cell suspension.
3. Mix vigorously by vertexing and incubate at 65 °C for 30 min, vortex the tube every 2–3 min to help release nucleic acids.
4. Centrifuge the mixture at 12,000× *g* for 15 min at room temperature.
5. Carefully transfer the supernatant to a new tube and add an equal volume of 8.0 M LiCl to the final concentration of 4.0 M (*see* **Note 18**).
6. Incubate at 4 °C for 8–18 h or at −20 °C for 1 h.
7. Centrifuge the mixture at 12,000× *g* for 30 min at 4 °C.
8. Discard the supernatant carefully and add 500 μL prechilled 70% ethanol.
9. Centrifuge at 7500× *g* for 5 min at 4 °C.
10. Repeat **steps 8** and **9** once.
11. Discard the supernatant carefully and add 500 μL prechilled 100% ethanol.
12. Centrifuge at 7500× *g* for 5 min at 4 °C and discard the supernatant carefully.

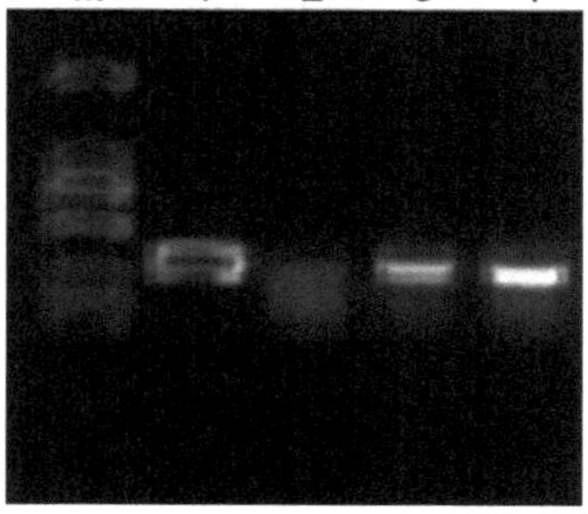

Fig. 2 Validation of dsRNA by nuclease treatment. Lanes 1–4 represent non-treated, DNase III-treated, DNase I-treated, and RNase A-treated dsRNA of GFP, respectively

13. Carefully eliminate the remaining ethanol by pipetting and air-dry the RNA pellets on a benchtop for 5 min (*see* **Note 19**).
14. Resuspend the RNA pellets in 180 μL of pre-warmed nuclease-free/DEPC-treated water, and take 4 μL for agarose gel electrophoresis, as in Fig. 2.
15. Add 20 μL 10×DNase buffer, 2 μL DNase I, and 2 μL RNase A, mix well by gently pipetting and incubate at 37 °C for 1 h to destroy remaining DNA and single-stranded RNA (*see* **Note 20**).
16. Store the dsRNA solution at −80 °C.

4 Notes

1. The empty vector is commercially available from Fisher Scientific under the Catalog No.69-864-3.
2. Chemically competent cells were used in this procedure; electrocompetent cells work as well.
3. Add 1% agar to make solid LB medium.
4. Sodium hydroxide is needed to prepare 0.5 M EDTA (pH8.0).
5. A 15-25 bp homologous arm included PCR primers for amplifying the target DNA fragment; also, the T7 promoter and lac operator were added to the forward primer for amplifying the pET-28a vector. Please refer to Fig. 1 for detailed design and primer sequences.
6. PCR was performed with High-fidelity DNA polymerases to avoid the introduction of potential mutation.

7. Excise the gel as small as possible and add at least 3 volumes Buffer GDP to 1 volume gel (100 mg gel~100 μL).
8. Make sure to add ethanol before use.
9. This step can be repeated once to increase the yield.
10. The optical amount of vector and insert fragment are 0.03 and 0.06 pmol, which can be calculated simply using the equations 0.02×base pair number of the vector and 0.04×base pair number of the fragment, respectively.
11. Do note heat-shock the bacteria longer than 1 min.
12. We routinely keep about 100 μL medium to resuspend the bacteria.
13. Follow the supplied protocol to purify the plasmid.
14. The O.D.600 value between 0.4 and 0.6 are all acceptable.
15. We find that IPTG concentration range from 0.1 to 3 mM had no significant influence on dsRNA production.
16. Over-cultivation cannot significantly improve the yield.
17. Avoid to generate air bubbles when resuspending the bacteria.
18. DsRNA is precipitated typically at 4.0 M LiCl. You can also use isopropanol to precipitate total nucleic acid (Equal volume of isopropanol is added to the aqueous layer and incubate at − 20 °C for at least 1 h).
19. Do not over-dry the RNA pellets, which may irreversibly reduce the solubleness of dsRNA.
20. We find that the degradation effect is better when the reaction time is appropriately extended.

References

1. Plasterk RHA (2002) RNA silencing: the genome's immune system. Science 296(5571): 1263–1265
2. Ding SW, Voinnet O (2007) Antiviral immunity directed by small RNAs. Cell 130(3): 413–426
3. de Felippes FF, Wang J-w, Weigel D (2012) MIGS: miRNA-induced gene silencing. Plant J 70(3):541–547
4. Senthil-Kumar M, Anand A, Uppalapati SR, Mysore KS (2008) Virus-induced gene silencing and its applications. CAB Rev 3
5. Smith NA et al (2000) Total silencing by intron-spliced hairpin RNAs. Nature 407(6802):319–320
6. Gan D et al (2010) Bacterially expressed dsRNA protects maize against SCMV infection. Plant Cell Rep 29(11):1261–1268
7. Yoon J et al (2021) Double-stranded RNA confers resistance to pepper mottle virus in Nicotiana benthamiana. Appl Biol Chem 64(1):1
8. McLoughlin AG et al (2018) Identification and application of exogenous dsRNA confers plant protection against Sclerotinia sclerotiorum and Botrytis cinerea. Sci Rep 8(1):7320
9. Liu S et al (2021) Evaluation of dsRNA delivery methods for targeting macrophage migration inhibitory factor MIF in RNAi-based aphid control. J Plant Dis Protect 128(5): 1201–1212
10. Kortmann M, Kuhl V, Klaffl S, Bott M (2015) A chromosomally encoded T7 RNA polymerase-dependent gene expression system for Corynebacterium glutamicum:

construction and comparative evaluation at the single-cell level. Microb Biotechnol 8(2): 253–265
11. Chalfie M, Tu Y, Euskirchen G, Ward W, Prasher D (1994) Green fluorescent protein as a marker for gene expression. Science 263(5148):802–805
12. Guan R, Chu D, Han X, Miao X, Li H (2021) Advances in the development of microbial double-stranded RNA production systems for application of RNA interference in agricultural pest control. Front Bioeng Biotechnol 9: 753790
13. Timmons LCDL, Fire A (2001) Ingestion of bacterially expressed dsRNAs can produce specific and potent genetic interference in Caenorhabditis elegans. Gene 263(1–2):103–112
14. Voloudakis AE et al (2015) Efficient double-stranded RNA production methods for utilization in plant virus control. Plant Virol Protocols 1236:255–274

Chapter 9

Inducible Expression of dsRNA in *Escherichia coli*

Saiya Duan and Guangjun Wang

Abstract

Double-stranded RNA (dsRNA) is a valuable tool for reverse genetics research and gene silencing applications. It is also an important management method for pests and diseases in agriculture. It can be synthesized both in vivo and in vitro. The latter presents the drawback of high production cost, the former is less expensive and suitable for scalable production. In general, dsRNAs are obtained in vivo from *Escherichia coli* heterologous systems that require the IPTG-inducible T7 RNA polymerase. In this report, we describe the construction of an RNAi system for the expression of dsRNA using the HT115 bacterial strain and the L4440 plasmid, and the extraction and identification of dsRNA.

Key words dsRNA synthesis, In vivo synthesis, *Escherichia coli*

1 Introduction

Since RNA interference was firstly identified in Cryptobacterium showyeri, it has been successfully applied to gene function analyses and gene therapy [6, 7, 18, 25], including plants, humans, insects, fungi, and nematodes [3, 17, 26]. In agriculture, RNA silencing has been considered a new and safe pest management strategy due to its high sequence-dependent specificity [5, 15]. Oral administration of double-stranded RNA (dsRNA) appears to be a convenient and practical approach for pest management. Studies have shown that transgenic plants expressing dsRNA have great potential for crop protection and have been applied to pest control [1, 14]. Currently a transdermal dsRNA delivery system for pest control has been established, which will facilitate the development of RNA pesticides [22, 23]. For plant pest and fungal disease control, the application of RNAi technology is also considered as a novel means of control [2, 11, 12], and opens new avenues for the control of post-harvest fungal diseases of fruits and vegetables [8]. Spray-induced gene silencing (SIGS) is a new RNAi-based method for plant protection, which uses in vitro synthesized dsRNA to spray the surface of fruits and vegetables that can be transported into

Xiaofei Cheng and Guanwei Wu (eds.), *Double-Stranded RNA: Methods and Protocols*, Methods in Molecular Biology, vol. 2771,
https://doi.org/10.1007/978-1-0716-3702-9_9,

pathogenic bacteria attached to the surface and then achieve transcriptional repression of target genes [9]. This method is easy to operate and does not require the host genetic transformation system. Moreover, the main component of this method is non-toxic and residue-free dsRNA, which should have broad application prospects in the field of plant disease control [21].

The preparation of dsRNA is usually done by transcribing and synthesizing the sense and antisense RNA sequences separately, and then annealing to produce dsRNA fragments, which is a complex, time-consuming, and costly process, and the quality of the obtained dsRNA is also not high. Although commercial kits have emerged, the high cost of dsRNA synthesis limits the large-scale application of dsRNA in the biological control of plant pests and diseases. Low-cost microbial reactors (e.g., *Escherichia coli*) are easy to operate and are able to rapidly and massively produce dsRNAs on inexpensive media. One of the most widely expressed systems is the production of dsRNAs in *E. coli* strain HT115 (DE3) [19, 20]. HT115 (DE3) is an RNaseIII-deficient strain that is unable to synthesize RNaseIII and therefore maintains dsRNA integrity, making it an ideal strain for large-scale preparation of dsRNAs [13]. Moreover, the synthesis of T7 RNA polymerase in HT115 (DE3) is controlled by the isopropyl-beta-D-thiogalactopyranoside (IPTG) induction, which in turn can strongly initiate the expression of RNAi vectors with the T7 promoter and synthesize large amounts of target dsRNA in a short time [16]. The L4440–HT115(DE3) system has been successfully applied for RNAi in several insects, including *Mythimna separate* [4], *Leptinotarsa decemlineata* [24], and *Bactrocera dorsalis* [10]. In this report, we present the detailed methods for the construction of the *E. coli* inducible expression systems, as well as steps for the identification and purification of dsRNA.

2 Materials

2.1 Equipment

1. Polymerase chain reaction (PCR) instrument.
2. Constant temperature oscillator (shaker).
3. Constant temperature incubator.
4. Centrifuge.
5. Erlenmeyer flasks and tubes.
6. Precision analytical digital lab scale.
7. Refrigerated Centrifuge.
8. pH meter.
9. Pipettes.
10. Autoclave.

11. Biosafety cabinet.
12. Centrifuge tubes.
13. NanoDrop UV/Vis Spectrophotometer (e.g., NanoDrop 8000, Thermo Scientific).
14. Gel electrophoresis system.

2.2 Chemical Reagents and Culture Medium

1. 100 mg/mL Ampicillin (Amp): Dissolve 1 g of Ampicillin sodium in 10 mL of sterile water in a biosafe cabinet, filtered through a 0.22 μm filter, dispensed into 1.5 mL sterilized centrifuge tubes and stored at −20 °C. The working concentration was 100 μg/mL, i.e., add 100 μL in 100 mL of medium.
2. 12.5 mg/mL Tetracycline (Tet): weigh 0.125 g of tetracycline hydrochloride powder on a precision laboratory scale and dissolve with 10 mL of anhydrous ethanol or 50% DMSO, filter through a 0.22 μm filter membrane and dispense into 1.5 mL sterile EP tubes and store at −20 °C. The working concentration is 12.5 ug/mL, i.e., add 100 μL in 100 mL of medium (*see* **Note 1**).
3. 1 M IPTG: dissolve 2.38 g IPTG in water, volume to 10 mL, filter through 0.22 μm filter membrane, dispense into 1.5 mL EP tubes and store at −20 °C. Use at a concentration of 1 mM/mL, i.e., add 100 μL in 100 mL of culture medium.
4. LB liquid medium: measure Tryptone 10 g, Yeast extract 5 g, NaCl 10 g, add deionized water, top up to 1 L with deionized water, and stir at room temperature until completely dissolved. Sterilize at 121 °C for 20 min.
5. LB solid medium: measure Tryptone 10 g, Yeast extract 5 g, NaCl 10 g, add about 700 mL deionized water, stir until completely dissolved, bring the total volume to 1 L with deionized water; divide into 200 mL each with 500 mL conical flasks, add 3 g agar to each flask, and steam sterilize for 20 min at 121 °C. Add suitable antibiotics to the medium before solidification (*see* **Note 2**).

2.3 Buffers

1. PBS buffer (pH 7.4) (1 L): NaCl 8 g, KCl 0.2 g, $Na_2HPO_4 \cdot 12H_2O$ 3.63 g, KH_2PO_4 0.24 g, adjust the pH to 7.4.
2. 75% Ethanol (diluted by PBS buffer).
3. 50 mM NaCl (autoclave at 121 °C for 20 min, and store at 4 °C).

2.4 Vector and Bacteria

1. L4440 vector.
2. RNase III-deficient *E. coli* HT115 (DE3) competent cells.

3 Methods

3.1 Interference Fragment Cloning and Interference Vector Construction

1. The L4440 vector contains an ampicillin resistance and a multiple cloning site (MCS) flanking with two T7 promoters (Fig. 1). Linear the L4440 vector by double cleavage with two restriction endonucleases, e.g., NheI and SalI. After incubating at 37 °C for 1 h, the reaction was then separated by electrophoresis in a 1% agarose gel, and the linearized L4440 was recovered with a gel DNA recovery kit (*see* **Note 3**).
2. Design primers for amplifying the interfering fragment of the gene of interest (GOI). The 5′ end of the forward primer and the 3′ end of the reverse primer contain the same restriction endonuclease sites (e.g., NheI and SalI) as for linearizing of L4440 vector (Fig. 2). After the thermal cycle, the PCR mixture was analyzed by electrophoresis on a 1% agarose gel, and the target DNA fragment was purified with a gel DNA recovery kit.
3. Insert the amplicon into pMD20-T vector according to the supplied instrument and transform into *E. coli* strain DH5α with standard molecular cloning procedure. Select the correct clones by PCR and extract the correct plasmid (e.g., pDM20-GOI) (*see* **Note 4**).

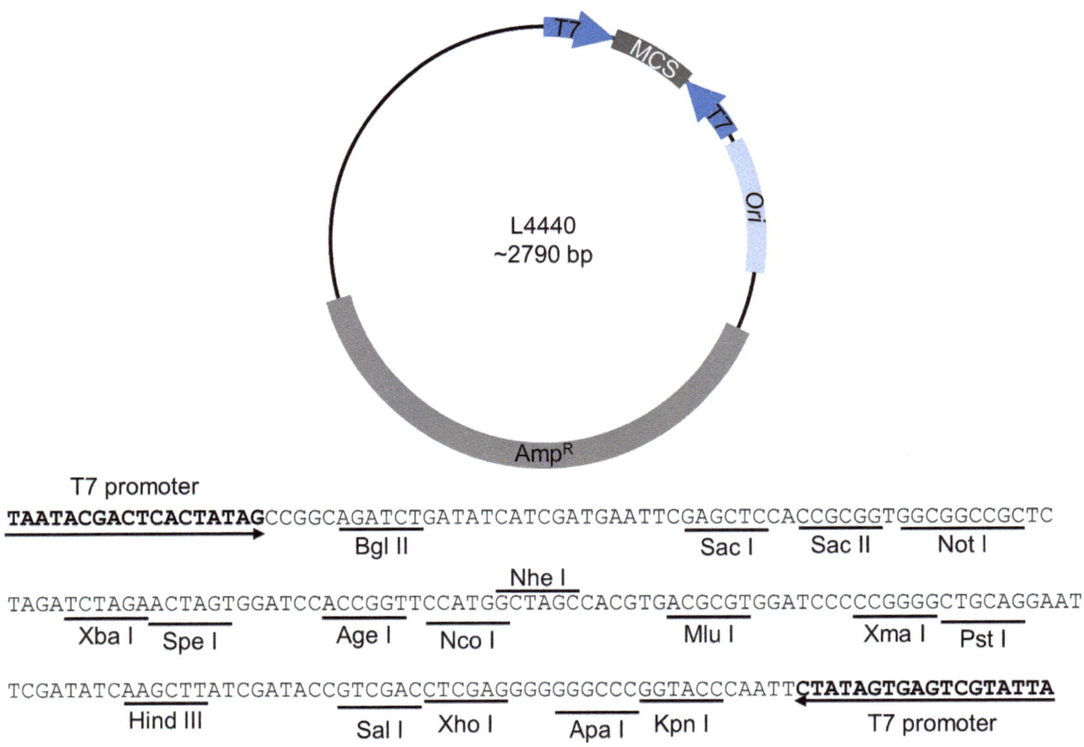

Fig. 1 Map of the L4440 vector

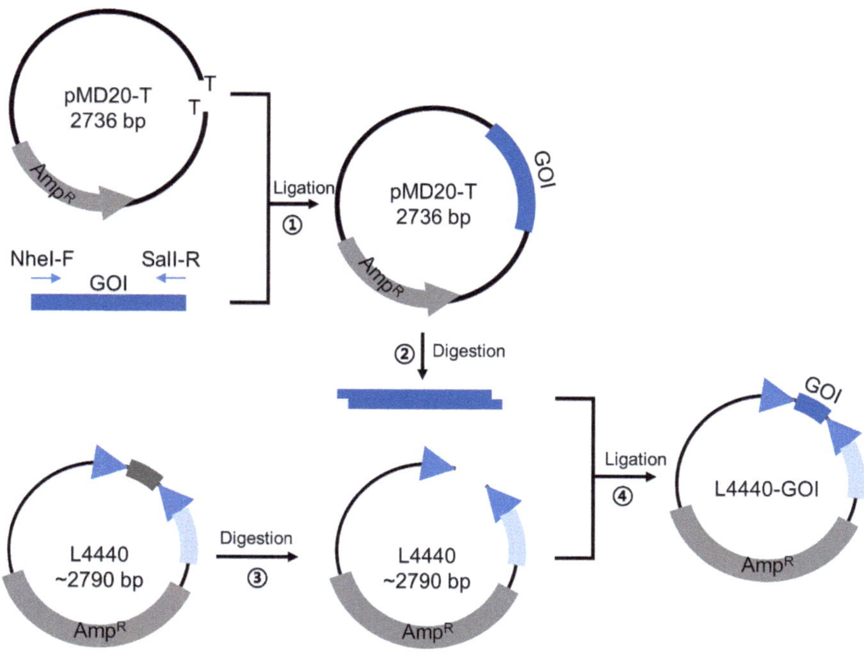

Fig. 2 Flow diagram of the construction of recombinant plasmid

4. Digest the pDM20-GOI by NheI and SalI, separated by electrophoresis in a 1% agarose gel, and recover the target DNA fragment with a gel DNA recovery kit.
5. Ligase linearized L4440 and the target DNA fragment from **step 4** using T4 DNA ligase and transform the recombinant plasmid (L4440-GOI) into *E. coli* strain DH5α with standard molecular cloning procedure. Select the correct clones by PCR, extract L4440-GOI plasmid, and then send the PCR-validated plasmids to the company for further Sanger sequencing (Fig. 2) (*see* **Note 5**).
6. Transform L4440-GOI into RNase III-deficient HT115 (DE3) as follows: thaw HT115 (DE3) competent cells on ice; add L4440-GOI plasmid (about 1 ng), mix well by gently pipetting, and incubate on ice for 30 min; place the HT115 (DE3) competent cells in a water bath at 42 °C for 90 s, transfer to ice immediately, and keep on ice for 2 min; add 800 μL of LB liquid medium and incubate at 200 rpm for 1 h at 37 °C (*see* **Note 6**).
7. Take about 200 μL LB medium from **step 6** and plate on LB plate supplemented with 100 μg/mL Amp and 50 μg/mL in a biosafe cabinet and incubate at 37 °C for 12~16 h. Positive clones were selected by PCR using universal primers SK-primer and KS-primer or other generic primers. Culture the correct clone in 5 mL of LB liquid medium containing 100 μg/mL

Amp and 50 μg/mL and incubate at 37 °C for 12~16 h in a shaking incubator. Add an equal volume of 30% glycerol to the bacterial culture and store the mixture at −80 °C for future use (*see* **Note 7**).

3.2 Induction of dsRNA Expression

1. Refresh the bacteria: take 200 μL bacterial culture of **step 7** of Subheading 3.1 in a biosafe cabinet, add 5 mL of LB liquid medium containing 100 μg/mL Amp and 50 μg/mL, and incubate at 37 °C overnight (12–16 h) at 200 rpm in a shaking incubator.
2. Growth of the bacteria: take 200 μL of the fresh bacterial culture from **step 1** and inoculate in a new 5 mL of liquid LB medium containing 100 μg/mL Amp and 50 μg/mL, and incubate at 37 °C for about 2 h until the OD_{600} value reaches about 0.6.
3. IPTG induction: Add 5 μL of IPTG to the bacterial broth and continue incubating for 4 h under the same conditions; the broth is now ready and contains the target dsRNA.

3.3 Extraction and Identification of dsRNA

The crude extraction of nucleic acids was carried out using the ethanol fraction method, with the following steps (*see* **Note 8**):

1. Centrifuge the IPTG-induced broth for 5 min at 4 °C at 8000 rpm and discard the supernatant.
2. Resuspend the bacterial pellets in 500 μL 75% ethanol by vortex and leave at room temperature for 5 min.
3. Centrifuge at 4 °C and 8000 rpm for 5 min and discard the supernatant.
4. Resuspend the bacterial pellets in 100 μL 50 mM NaCl solution and leave to stand at room temperature for 1 h.
5. After centrifugation at 4 °C and 8000 rpm for 10 min, the supernatant containing the crude nucleic acids was transferred to a new centrifuge tube.
6. DNase I treatment: prepare the enzymatic reaction as follows: 1.0 μL 10× DNase I buffer, 1.0 μL DNase I (1 U/μL), 1.0 μL crude nucleic acids, 7.0 μL ddH_2O, mix well by pipetting, and incubate at 37 °C for 30 min;
7. RNase A treatment: add 0.75 μL RNase A (1.0 μg/mL), 4.25 μL 1 mmol/L NaCl, 4.0 μL ddH_2O, 1.0 μL crude nucleic acids, mix well by pipetting, and incubate at 37 °C for 1 h;
8. After the DNase I and RNase A treatment, analyze the reaction mixtures by 1% agarose gel electrophoresis and the buffer should be DEPC-treated (Fig. 3) (*see* **Note 9**).

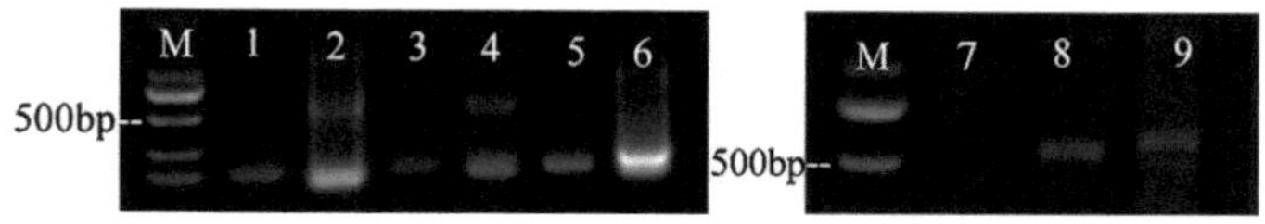

Fig. 3 Agar gel electrophoresis of dsRNA. (**a**) The crude nucleic acid extract that has not been treated by DNase I and RNase A; (**b**) the crude nucleic acid extract that has been treated by DNase I and RNase A; Lanes 1, 3, and 5 are crude nucleic acid extracts of uninduced bacteria; Lanes 2, 4, 6, 7, 8, and 9 are crude extracts of IPTG-induced bacteria; Lanes 5, 6, and 7 are empty vector as control

4 Notes

1. Screw cap tubes are not recommended for Amp and Tet dispensing as they are prone to bring contamination.
2. When preparing the resistance medium, ensure that all of the antibiotic is added to the medium. Solid media should not be reheated and melted if antibiotics have been added to prevent the weakening of resistance.
3. It is not recommended to select two restriction enzyme sites next to each other to prevent incomplete cleavage.
4. If seamless ligation is chosen, the primers are designed to remove the enzyme cut site and at least 12 bases of homologous sequence should be added at each end for recombination.
5. After double digestion, the digested product can be subjected to gel recovery to increase the efficiency of subsequent ligation.
6. Reduce the recovery time or the amount of bacterial solution to be plated, if there are too many positive clones.
7. At least 15% glycerol should be added to the bacterial culture to protect the bacteria. If the bacterial culture is stored for a long period, it should be verified by sequencing before use.
8. Other RNA extraction methods can also be used.
9. Inadequate enzymatic digestion may cause extra bands in the gel. Increasing the digestion time can reduce the problem.

References

1. Baum JA, Bogaert T, Clinton W, Heck GR, Feldmann P, Ilagan O, Roberts J (2007) Control of coleopteran insect pests through RNA interference. Nat Biotechnol 25(11): 1322–1326
2. Cagliari D, Dias NP, Galdeano DM, Dos Santos EÁ, Smagghe G, Zotti MJ (2019) Management of pest insects and plant diseases by non-transformative RNAi. Front Plant Sci 10: 1319
3. Dubrovina AS, Kiselev KV (2019) Exogenous RNAs for gene regulation and plant resistance. Int J Mol Sci 20(9):2282
4. Ganbaatar O, Cao B, Zhang Y, Bao D, Bao W (2017) Knockdown of Mythimna separata chitinase genes via bacterial expression and oral delivery of RNAi effectors. BMC Biotechnol 17(1):9

5. Gordon KH, Waterhouse PM (2007) RNAi for insect-proof plants. Nat Biotechnol 25(11): 1231–1232
6. Kamath RS, Fraser AG, Dong Y, Poulin G, Durbin R, Gotta M, Ahringer J (2003) Systematic functional analysis of the *Caenorhabditis elegans* genome using RNAi. Nature 421(6920):231–237
7. Kim J, Badaloni A, Willert T, Zimber-Strobl U, Kühn R, Wurst W, Kieslinger M (2013) An RNAi-based approach to down-regulate a gene family in vivo. PLoS One 8(11):e80312
8. Koch A, Biedenkopf D, Furch A, Weber L, Rossbach O, Abdellatef E, Kogel KH (2016) An RNAi-based control of *Fusarium graminearum* infections through spraying of long dsRNAs involves a plant passage and is controlled by the fungal silencing machinery. PLoS Pathog 12(10):e1005901
9. Koch A, Höfle L, Werner BT, Imani J, Schmidt A, Jelonek L, Kogel KH (2019) SIGS vs HIGS: a study on the efficacy of two dsRNA delivery strategies to silence *Fusarium* FgCYP51 genes in infected host and non-host plants. Mol Plant Pathol 20(12):1636–1644
10. Li X, Zhang M, Zhang H (2011) RNA interference of four genes in adult *Bactrocera dorsalis* by feeding their dsRNAs. PLoS One 6(3): e17788
11. Liu S, Jaouannet M, Dempsey DMA, Imani J, Coustau C, Kogel KH (2020) RNA-based technologies for insect control in plant production. Biotechnol Adv 39:107463
12. Mamta B, Rajam MV (2017) RNAi technology: a new platform for crop pest control. Physiol Mol Biol Plants 23:487–501
13. Niño-Sánchez J, Chen LH, De Souza JT, Mosquera S, Stergiopoulos I (2021) Targeted delivery of gene silencing in fungi using genetically engineered bacteria. J Fungi 7(2):125
14. Pitino M, Coleman AD, Maffei ME, Ridout CJ, Hogenhout SA (2011) Silencing of aphid genes by dsRNA feeding from plants. PLoS One 6(10):e25709
15. Price DR, Gatehouse JA (2008) RNAi-mediated crop protection against insects. Trends Biotechnol 26(7):393–400
16. Rio DC (2013) Expression and purification of active recombinant T7 RNA polymerase from E coli. Cold Spring Harbor Protocols 2013(11)
17. Silva J, Chang K, Hannon GJ, Rivas FV (2004) RNA-interference-based functional genomics in mammalian cells: reverse genetics coming of age. Oncogene 23(51):8401–8409
18. Takahashi Y, Nishikawa M, Takakura Y (2009) Nonviral vector-mediated RNA interference: its gene silencing characteristics and important factors to achieve RNAi-based gene therapy. Adv Drug Deliv Rev 61(9):760–766
19. Tenllado F, Martínez-García B, Vargas M, Díaz-Ruíz JR (2003) Crude extracts of bacterially expressed dsRNA can be used to protect plants against virus infections. BMC Biotechnol 3:3
20. Timmons L, Court DL, Fire A (2001) Ingestion of bacterially expressed dsRNAs can produce specifc and potent genetic interference in *Caenorhabditis elegans*. Gene 263:103–112
21. Wang M, Thomas N, Jin H (2017) Cross-kingdom RNA trafficking and environmental RNAi for powerful innovative pre- and post-harvest plant protection. Curr Opin Plant Biol 38:133–141
22. Yan S, Qian J, Cai C, Ma Z, Li J, Yin M, Shen J (2020) Spray method application of transdermal dsRNA delivery system for efficient gene silencing and pest control on soybean aphid Aphis glycines. J Pest Sci 93:449–459
23. Zheng Y, Hu Y, Yan S, Zhou H, Song D, Yin M, Shen J (2019) A polymer/detergent formulation improves dsRNA penetration through the body wall and RNAi-induced mortality in the soybean aphid Aphis glycines. Pest Manag Sci 75(7):1993–1999
24. Zhu F, Xu J, Palli R, Ferguson J, Palli SR (2011) Ingested RNA interference for managing the populations of the Colorado potato beetle, *Leptinotarsa decemlineata*. Pest Manag Sci 67(2)
25. Zotti MJ, Smagghe G (2015) RNAi technology for insect management and protection of beneficial insects from diseases: lessons, challenges and risk assessments. Neotrop Entomol 44:197–213
26. Zotti M, Dos Santos EA, Cagliari D, Christiaens O, Taning CNT, Smagghe G (2018) RNA interference technology in crop protection against arthropod pests, pathogens and nematodes. Pest Manag Sci 74(6): 1239–1250

Chapter 10

In Vivo Production of dsRNA Using Bacteriophage ϕ6 in *Pseudomonas syringae* Cit7 Cells

Jinguang Yang

Abstract

RNA interference (RNAi), also known as post-transcriptional gene silencing (PTGS), is one of the emerging genetic engineering techniques to effectively silence or inhibit the expression of target genes. This chapter describes a method for in vivo production of dsRNA in non-pathogenic *Pseudomonas syringae* strains using phage ϕ6 RNA-dependent RNA polymerase, extraction and purification of dsRNA from bacterial solution, and the use of dsRNA to induce silencing of green fluorescent protein (GFP) in transgenic *Nicotiana benthamiana*.

Key words RNAi, dsRNA, Bacteriophage ϕ6, Absorption, Crop protection

1 Introduction

RNA interference (RNAi), also known as post-transcriptional gene silencing (PTGS) or RNA silencing (RNAi), is a conserved mechanism for regulating gene expression in eukaryotes. Double-stranded RNA (dsRNA) is a key component in inducing RNAi. During the initiation of RNAi, dsRNA molecules are recognized and processed by Dicer or Dicer-like endonucleases into short interfering RNA (siRNA), which then binds to the ribozyme complex caused RNA-induced silencing complex (RISC), which contains the key protein AGO (Argonaute), for subsequent gene expression regulation [9]. Exogenous spray of dsRNA can also induce the silencing of endogenous gene, which has important practice potential for disease and pest control in agriculture. However, the mass-scale synthesis of dsRNA with low cost is the major challenge of this technology. In addition to the challenges of massive production of dsRNA, there are also challenges in the delivery of dsRNA for biocontrol and RNAi efficiency. Nevertheless, it is

Xiaofei Cheng and Guanwei Wu (eds.), *Double-Stranded RNA: Methods and Protocols*, Methods in Molecular Biology, vol. 2771, https://doi.org/10.1007/978-1-0716-3702-9_10,

expected that RNAi-based biopesticides will enter the market for foliar application, trunk injection, root dipping, or seed treatment in the near future [14, 15].

At present, both in vivo and in vitro synthesis of dsRNA mainly rely on the bacteriophage T7 DNA-dependent RNA polymerase (DdRp) and its promoter for transcription for specific sequences in *RNase* III-defective *Escherichia coli HT115* (DE3) [2, 7]. Recently, large-scale dsRNA production systems were established in non-pathogenic *P. syringae* Cit7 cells expressing the bacteriophage ϕ6 RNA-dependent RNA polymerase (RdRP), and the genomic fragment of ϕ6 that has been inserted with the target gene was used to establish a stable carrier state relationship with the host cells for producing dsRNA [3–6]. The bacteriophage ϕ6 expression system uses ϕ6 RdRP to synthesize complementary RNA strands from the ssRNA template to generate dsRNA. Basically, this system needs a plasmid to express T7 DdRp for generating the ssRNA template and a plasmid to express phi 6 RdRp to synthesize dsRNA from the ssRNA template. This system also needs the third plasmid, which contains the target sequence in the phi 6 genomic sequence coding the precapsid protein under the T7 promoter. Compared with the T7 system, the ϕ6 expression system produces dsRNA much more efficiently.

Bacteriophage ϕ6 is the representative member of the family *Cystoviridae.* Bacteriophage ϕ6 has a tri-segmented double-stranded RNA (dsRNA) genome of about 13.5 kb that is encapsulated in enveloped, double-capsid spherical particles. Bacteriophage ϕ6 was first isolated from the plant pathogenic *Pseudomonas phaseolicola* that comes from diseased legumes by Vidaver et al. in 1973 [1]. Bacteriophage ϕ6 can infect non-plant pathogenic *P. adaceae* as well. The genome of bacteriophage ϕ6 consists of three dsRNA segments, that are named based on the length as S, M, and L, respectively (Fig. 1). The L segment is about 6374 bp in length and encodes four phage-specific proteins (P1, P2, P4, and P7), which constitute the viral replication complex, which is also called protocapsid (PC), and is required for infecting the plant pathogen *P. phaseolicola* [8]. The M segment is about 4063 bp in length and encodes four viral proteins (P3, P6, P10, and P11), and the S segment is about 2948 bp in length and encodes four proteins (P5, P8, P9, and P12) [8]. In the ϕ6 infection cycle, PCs envelope the single-stranded genome precursors and convert it into dsRNA within the PC. The viral dsRNAs are never exposed to cytoplasm. P2 within the PCs transcribe mRNA from dsRNA and export them to the cytoplasm for protein synthesis.

P2 is the viral RNA-dependent RNA polymerase (RdRP) and can be used to produce large quantities of high-quality dsRNA in vitro. Synthesizing full-length dsRNA by combining T7 RNA polymerase transcription and ϕ6 RdRP has virtually no length

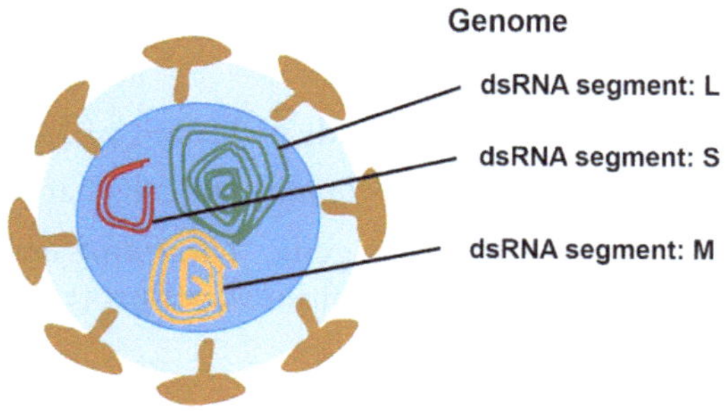

Fig. 1 Genome structure of bacteriophage ϕ6

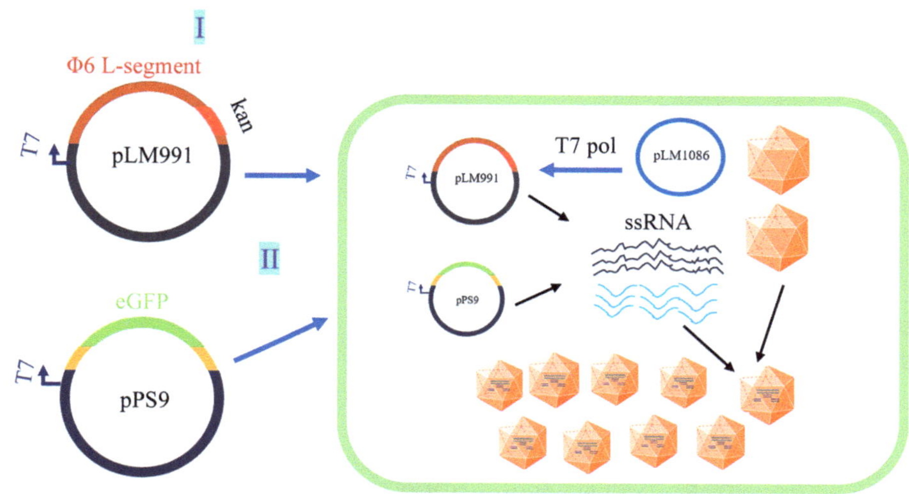

Fig. 2 dsRNA production system using phage ϕ6

limitation. Bacteriophage ϕ6 is also an excellent tool for large-scale synthesis of short and medium dsRNA in the carrier status of *P. phaseolicola* HB10Y [4, 10]. The carrier status of bacterial cells can be obtained by inserting the genome fragment L into a ColEI-based suicide plasmid controlled by a T7 promoter in the non-pathogenic *P. phaseolicola* Cit7 cells [5]. A kanamycin resistance gene was inserted into 3′ untranslated regions of the L fragment for the selection of stable carrier state cells. The *P. phaseolicola* Cit7 cells also harbor a plasmid (pLM1086) that constitutionally expresses the T7 RNA polymerase [5]. The expression of T7 RNA polymerase enables the expression of the four viral proteins to assemble PCs in the non-pathogenic *P. phaseolicola* Cit7 cells. Plasmids containing T7 promoters and viral cDNA are transcribed and packaged into viral PC particles, which then synthesize complementary RNA strands to further synthesize dsRNA (Fig. 2) [11, 12].

2 Materials

2.1 Plasmids, Bacteria, and Plant Materials

1. Plasmids pLM659, pLM991, and pPS9.
2. *P. syringae* Cit7 cells harboring the pLM1086 plasmid.
3. Transgenic *Nicotiana benthamiana* seedlings expressing the green fluorescent protein (GFP).

2.2 Equipment

1. Laser confocal microscope.
2. Centrifuge.
3. Thermal cycler.
4. Incubator.
5. Shaker.
6. Flask.
7. Pipettes.
8. Biosafety cabinet.
9. Gel electrophoresis system.
10. Electroporation device.

2.3 Chemical and Reagents

1. Kanamycin.
2. Kings' B medium.
3. LB liquid medium.
4. Sterile water.
5. 2×STE solution (200 mM NaCl, 0.1 M Tris–HCl pH 8.0, 2 mM EDTA).
6. Phenol/chloroform/isopentyl alcohol (25:24:1).
7. β-mercaptoethanol.
8. Sodium dodecyl sulfate (SDS).
9. Anhydrous ethanol.
10. CF-11 cellulose column.
11. Isopropanol.
12. RNase-Free DNase.
13. PBS.
14. Agarose.
15. Nucleic acid dyes.
16. Syringes.

3 Methods

3.1 Plasmid Construction

Plasmid pLM659 contains a cDNA copy of the 2948-bp S segment of the bacteriophage ϕ6 in the pT7T319U backbone [13]. Plasmid pLM991 contains a 7599-bp cDNA copy of the L segment of the bacteriophage ϕ6 and the 3′ end contains a kanamycin resistance gene [5]. Plasmid pPS9 was derived from pLM659 and contains the eGFP gene flanked by 5′ and 3′ regulatory sequences from the bacteriophage ϕ6 S segment. pPS9 was constructed by amplifying the partial fragment of *GFP* gene by PCR from the Fu28-GFP plasmid using primers 5′-CGGCCGGCCGAGCTCGAATTCGG TACCATATGGTGAGCAAGGGCGAGGAGCTGTTC-3′ and 5′- -CGGC*CTGCAG* ATCTTAAGTTTAAACCTTGTA CAGCTCGTCCATGCCGAGAGTGA-3′ that contain restriction sites of *Eag*I (underlined) and *Pst*I (underlined and italic) (*see* **Note 1**). The *GFP* fragment and pLM659 were digested with *Eag*I and *Pst*I and subsequently ligated together by T4 DNA ligase to yield pPS9 [3]. The ligation product was transformed into pPS9 *E. coli* with normal cloning processes, plasmid was extracted with a mini-prep plasmid purification kit, and was confirmed by Sanger sequencing (*see* **Note 2**).

3.2 Electroporation

Equal amount of pLM659 and pPS9 were mixed with electroporation-competent *P. syringae* Cit7 cells, mixed well gently, and incubated on ice for ten minutes. The mixture was added into the electroporation cuvette and electroporated at 1500 voltage. About 150 μL LB liquid medium was added into the bacteria and the mixture was incubated on a shaker incubator at 37 °C for 1 h. The mixture was plated on solid LB medium and the plate was incubated at 37 °C for 48 h.

3.3 In Vivo dsRNA Production

Cultures of *P. syringae* Cit7 (pLM1086) cells transfected with pLM659 were cultured in Kings' B medium at 28 °C for 24 h to induce the synthesis of dsRNA (*see* **Note 3**). The cultures were frozen on ice for 15 min at 4 °C, centrifuged at 8000 g for 10 min, and the bacterial pellets were resuspended in 10 mL of sterile water and lysed by ultrasonication for 1 h on ice.

3.4 Extraction of dsRNA

Add 500 μL of 2×STE solution (pH 8.0), 500 μL of phenol/chloroform/isoamyl alcohol (25:24:1), 15 μL of β-mercaptoethanol, and 50 μL of 10% SDS to 1 mL of ultrasonic wave-treated bacterial solution. Mix well by vertexing for 3 min, and centrifuge at 8000 r/min for 10 min. The supernatant was adjusted to a final concentration of 17% ethanol with anhydrous ethanol and added to a CF-11 cellulose column. After washing the column with 10 volumes of 1×STE solution containing 17% ethanol, the column was eluted with 1×STE solution without ethanol.

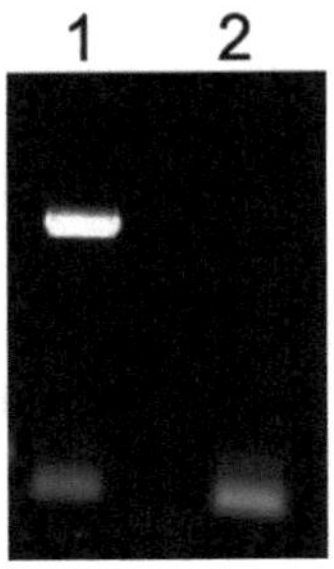

Fig. 3 Agarose gel analysis of dsRNA produced by the in vivo production system. Lane 1: dsRNA from GFP carrier state cells extracted by CF-11 cellulose column, lane 2: dsRNA control

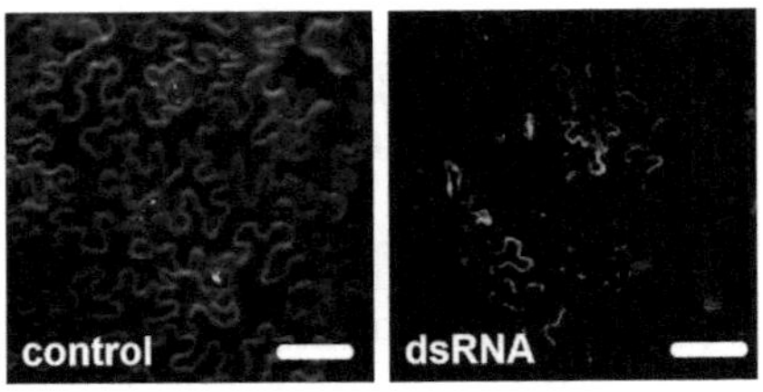

Fig. 4 Representative confocal images of *N. benthamiana* leaves infiltrated by dsRNA at 3 dpi. Scale bar, 50 μm

The nucleic acid eluate was detected and collected under a nucleic acid UV detector (*see* **Note 4**). The eluate was added to an equal volume of pre-cooled isopropanol and incubated for 30 min at −20 °C. The dsRNAs were precipitated by centrifuge at 12,000 rpm for 20 min at 4 °C. The dsRNA pellets were washed with 75% ethanol for 15 min at 12,000 r/min and dissolved in 30 μL of RNase-free sterile water and stored at −20 °C (*see* **Note 5**). A small fraction (5 μL) of the sample was analyzed by gel electrophoresis (Fig. 3).

3.5 Delivery of dsRNA into Plant Leaves

Grow the transgenic *N. benthamiana* plants to the right size (about 4 weeks old). Dilute the extracted dsRNA solution with PBS into 50 mM and about 500 μL of the dsRNA solution was infiltrated into the third leaf of the *N. benthamiana* seedlings using a 1-mL needle-less syringe (*see* **Note 6**). At 3 days post-infiltration (dpi), the transcript level of *GFP* can be analyzed by Northern blot and the accumulation of GFP protein can be detected by western blot with antibodies against GFP or directly analyzed by a confocal microscope (Fig. 4).

4 Notes

1. Other target DNA fragment can be amplified using primers that were designed with similar strategy.
2. Ensure that the vector digestion process is complete, otherwise false positives may occur. If incomplete digestion occurs, the length of digestion and the amount of restriction enzyme can be adjusted appropriately. The target gene should also be added proportionally when ligating to the vector to ensure the efficiency of vector ligation.
3. Increasing the expression time may reduce the overall dsRNA yield due to the exhaustion of nutrition in the medium.
4. To improve the extraction efficiency of dsRNA, CF-11 cellulose column can be rinsed more than once to improve the elution efficiency.
5. Make sure all solutions and devices are RNase-free or treated by DEPC to reduce the degradation of dsRNA.
6. Inject *N. benthamiana* leaves slowly and gently to avoid damaging the tissue.

References

1. Vidaver AK, Koski RK, Van Etten JL (1973) Bacteriophage phi6: a lipid-containing virus of Pseudomonas phaseolicola. J Virol 1973:799–805. https://doi.org/10.1128/JVI.11.5.799-805.1973
2. Das PR, Sherif SM (2020) Application of exogenous dsRNAs-induced RNAi in agriculture: challenges and triumphs. Front Plant Sci 2020:946. https://doi.org/10.3389/fpls.2020.00946
3. Aalto AP, Sarin LP, van Dijk AA, Saarma M, Poranen MM, Arumae U, Bamford DH (2007) Large-scale production of dsRNA and siRNA pools for RNA interference utilizing bacteriophage phi6 RNA-dependent RNA polymerase. RNA 2007:422–429. https://doi.org/10.1261/rna.348307
4. Onodera S, Olkkonen VM, Gottlieb P, Strassman J, Qiao XY, Bamford DH, Mindich L (1992) Construction of a transducing virus from double-stranded RNA bacteriophage phi6: establishment of carrier states in host cells. J Virol 1992:190–196. https://doi.org/10.1128/JVI.66.1.190-196.1992
5. Sun Y, Qiao X, Mindich L (2004) Construction of carrier state viruses with partial genomes of the segmented dsRNA bacteriophages. Virology 2004:274–279. https://doi.org/10.1016/j.virol.2003.10.022
6. Wilson M, Hirano SS, Lindow SE (1999) Location and survival of leaf-associated bacteria in relation to pathogenicity and potential for growth within the leaf. Appl Environ Microbiol 1999:1435–1443. https://doi.org/10.1128/AEM.65.4.1435-1443.1999
7. Timmons L, Court DL, Fire A (2001) Ingestion of bacterially expressed dsRNAs can produce specific and potent genetic interference in Caenorhabditis elegans. Gene 2001:103–112. https://doi.org/10.1016/s0378-1119(00)00579-5
8. Mindich L, Nemhauser I, Gottlieb P, Romantschuk M, Carton J, Frucht S, Strassman J, Bamford DH, Kalkkinen N (1988) Nucleotide sequence of the large double-stranded RNA segment of bacteriophage phi 6: genes specifying the viral replicase and transcriptase. J Virol 1988:1180–1185. https://doi.org/10.1128/JVI.62.4.1180-1185.1988
9. Carthew RW, Sontheimer EJ (2009) Origins and mechanisms of miRNAs and siRNAs. Cell 2009:642–655. https://doi.org/10.1016/j.cell.2009.01.035

10. Poranen MM, Tuma R (2004) Self-assembly of double-stranded RNA bacteriophages. Virus Res 2004:93–100. https://doi.org/10.1016/j.virusres.2003.12.009
11. Levanova A, Poranen MM (2018) RNA interference as a prospective tool for the control of human viral infections. Front Microbiol 2018:2151. https://doi.org/10.3389/fmicb.2018.02151
12. Niehl A, Soininen M, Poranen MM, Heinlein M (2018) Synthetic biology approach for plant protection using dsRNA. Plant Biotechnol J 2018:1679–1687. https://doi.org/10.1111/pbi.12904
13. Gottlieb P, Strassman J, Qiao X, Frilander M, Frucht A, Mindich L (1992) In vitro packaging and replication of individual genomic segments of bacteriophage phi 6 RNA. J Virol 1992:2611–2616. https://doi.org/10.1128/JVI.66.5.2611-2616.1992
14. Zotti MJ, Smagghe G (2015) RNAi technology for insect management and protection of beneficial insects from diseases: lessons, challenges and risk assessments. Neotrop Entomol 2015:197–213. https://doi.org/10.1007/s13744-015-0291-8
15. San MK, Scott JG (2016) The next generation of insecticides: dsRNA is stable as a foliar-applied insecticide. Pest Manag Sci 2016:801–809. https://doi.org/10.1002/ps.4056

Chapter 11

Bacteria-Based Double-Stranded RNA Production to Develop Cost-Effective RNA Interference Application for Insect Pest Management

Ruobing Guan, Xuexia Miao, and Haichao Li

Abstract

Since the discovery of the RNA interference (RNAi) mechanism, it has been widely used in the fields of gene function, biomedicine, and crop pest control. In the direction of agricultural application, this technology is highly expected, especially in the field of pest control, which is called "the third revolution in the history of pesticides". Currently, RNA biopesticides are developing rapidly all over the world. A genetically modified product (MON87411) has been approved for marketing, and a large number of agricultural companies are developing products based on direct spraying RNA biopesticides and submitting them for regulatory approval. The biggest problem that has emerged for spray RNA biopesticides is the technology for large-scale and low-cost production of dsRNA. At present, the bacterial fermentation production technology can realize large-scale dsRNA production with a yield of 4.23~182 mg/L bacterial solution. This article describes the experimental protocol for dsRNA production technology based on bacterial fermentation.

Key words RNA interference, Double-stranded RNA, Bacteria-based production, RNAi pesticides

1 Introduction

RNA biopesticides use the principle of RNA interference (RNA interference, RNAi) to inhibit the expression of important genes of target organisms, thereby achieving the purpose of pest control. Due to the characteristics of strong targeting, high efficiency, green, and pollution-free, RNA biopesticides are called "the third revolution in the history of pesticides" [1]. RNAi technology can be used to design species-specific RNA biopesticides for certain diseases or pests without affecting non-target organisms; it is also possible to design interspecies broad-spectrum RNA biopesticides targeting multiple species by using shared targets between species; formulations with dsRNA as the main form can be rapidly degraded in the environment. Under the premise of ensuring control efficiency, the problems of residue and environmental pollution are

Xiaofei Cheng and Guanwei Wu (eds.), *Double-Stranded RNA: Methods and Protocols*, Methods in Molecular Biology, vol. 2771, https://doi.org/10.1007/978-1-0716-3702-9_11,

almost negligible; this technology only temporarily shuts down the expression of a certain gene of the pest, and does not change the genome of the organism itself, which does not produce heritable mutations, so it almost does not affect the ecosystem; the target gene is highly replaceable and not easy to produce resistance; at the same time, RNA biological pesticide products have incomparable advantages over chemical pesticides and conventional microbial pesticides, such as high purity, quick effect, non-toxicity and pollution-free, wide prevention and control range and low price, and have obvious advantages in market competition [5, 12].

However, the biggest problem affecting the application of RNAi technology in the agricultural field is the low-cost and large-scale production system of dsRNA. With the continuous efforts of researchers, the production cost of dsRNA has dropped from 12,000 US dollars per gram in 2008 to 1 US dollar per gram in 2021 [4, 11]. Normally, the amount of dsRNA used is very low, only 1~5 g per acre (https://www.forbes.com/sites/michaelhelmstetter/2020/08/06/rnai-based-pesticides-contribute-to-the-promise-of-a-new-green-revolution/?sh=7812b97178e7). From this perspective, the current production cost of dsRNA is perfectly adequate for commercial applications. Solving the price problem of RNA biopesticide indicates that the fundamental problem of commercial application of RNA biopesticide has been effectively solved.

At the same time, there are many options for producing dsRNA, and various solutions such as bacterial expression, yeast expression, and cell-free synthesis can realize the large-scale production of dsRNA. For example, Greenlight Biosciences has established a variety of platforms for dsRNA production processes. Their cell-free production systems can control the production cost of dsRNA at a price of 1 US dollar per gram , and can guarantee the production of technical grade active ingredients (TGAI) of dsRNA. The production system can realize the production size of dsRNA from 200 to 4000 bp, and realize the production yield scale from mg to kg scale. RNAgri's dsRNA synthesis platform can control the production cost of dsRNA at 1 US dollar per gram, and Genolution can synthesize milligram (mg) to kilogram (kg) scale dsRNA (200–800 bp) through a fermentation platform (http://genolution.co.kr/). Renaissance BioScience has also established a platform for yeast fermentation to produce dsRNA, which can be used for pest control by expressing dsRNA for different targets in yeast.

The yield of dsRNA depends on the expression plasmid and strain. The most commonly used production system is HT115-L4440, which produces dsRNA under the control of an inducible T7 promoter (Fig. 1). Vectors used for the expression of dsRNA typically contain single or double T7 promoters, and constructs containing a single T7 promoter (such as pGEM-T and pET28a/22b) appear to produce better results than constructs containing

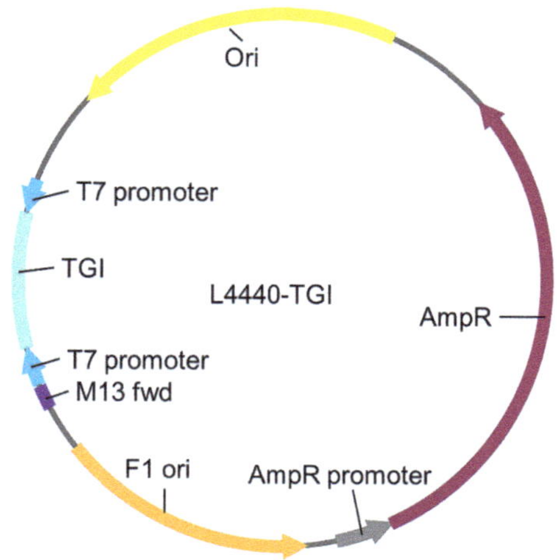

Fig. 1 Construction of bacterial expression vectors for dsRNA production

double T7 promoters (such as pL4440). The high dsRNA yields [8, 13] may be due to the different energy consumed by the single and double T7 promoters during transcription. In addition, the RNA transcribed from a single T7 promoter contains two reverse complementary strands flanked by a loop sequence, and this structure is more easily annealed, whereas the double T7 promoter allows bidirectional transcription of the insert, and the transcribed ssRNA is more easily degraded by RNA endonucleases before forming dsRNA.

The yield of dsRNA is also affected by the host strain, for example, RNaseIII-deficient strains of M-JM109 and M-JM109 lacY produced higher dsRNA yields than HT115 (DE3) strains [13]. The pET28-BL21 (DE3) RNase III-system produced a higher amount of dsRNA than the pET28-HT115 (DE3) system, while the dsRNA expression efficiency of the L4440-BL21 (DE3) RNase III-system was lower than that of the L4440-HT115 (DE3) [8]. Factors affecting dsRNA expression efficiency may be due to suboptimal dsRNA production conditions, nutrition, and fermentation process [4, 9]. At present, the dsRNA yield achieved by this system is 4.23~182 mg/L bacterial liquid, which is much higher than the traditional production, but the yield still needs to be greatly improved [8, 10]. Meanwhile, it is well known that IPTG is a commonly used reagent to induce dsRNA expression in vector systems containing T7 promoter, but due to the cost and toxicity of IPTG [3], it is gradually replaced by lactose and skimmed milk to produce dsRNA [2, 7, 9, 10]. These natural materials are inexpensive and, therefore, have great potential for large-scale production of dsRNA. In addition, a constitutive expression system that expresses dsRNA without the addition of inducers is also a very promising expression system [2, 6]. Based on this, this chapter introduces the detailed process of producing dsRNA based on bacterial fermentation.

2 Materials

All materials need to be prepared in advance, L4440 plasmid, HT115 (DE3) strain, etc. need to be activated and tested in advance, the culture medium is sterile and all reagents are prepared with sterile water. Prepare the following kits and reagents in advance: Total RNA extraction kit Trizol, cDNA first-strand synthesis kit, SYBR Premix Ex Taq real-time fluorescent quantitative PCR kit, DNA Polymerase, T4 DNA ligase, DNase I, RNaseA, restriction endonuclease, ampicillin (Amp), X-gal, IPTG, tetracycline hydrochloride (Tet), DNA recovery kit, small plasmid extraction kit, isopropanol, chloroform, absolute ethanol and other analytical reagents.

2.1 Preparation of LB Liquid Medium

1. Tryptone 10 g/L, Yeast extract 5 g/L, NaCl 10 g/L, adjust the pH to 7.0 with 5 mol/L NaOH, and dilute to 1 L with sterile water (*see* **Note 1**).
2. Stirring at room temperature with a magnetic stirrer for 30 min.
3. Steam sterilize with an autoclave at 15 psi, 121 °C for 20 min, take it out for later use.

2.2 Preparation of LB Solid Medium

1. Tryptone 10 g/L, Yeast extract 5 g/L, NaCl 10 g/L, Agar 15 g/L, dilute to 1 L with sterile water (*see* **Note 1**).
2. Stirring at room temperature for 30 min using a magnetic stirrer.
3. Steam sterilize at 15 psi, 121 °C for 20 min using an autoclave.
4. Melt the LB solid medium by heating and add antibiotics when the temperature of the medium drops to 55 °C and shake well.
5. Pour the LB medium evenly into a sterile 9 cm Petri dish, seal it with a parafilm after cooling, store at 4 °C, and use it within 1 month (*see* **Note 2**).

2.3 Preparation of Competent Cells

1. Select HT115(DE3) single clones and inoculate them into 5 mL LB liquid medium containing 12.5 μg/mL Tet+, and culture overnight at 37 °C with shaking (*see* **Note 11**).
2. Inoculate 1 mL of bacterial broth into 100 mL of LB liquid medium containing Tet+, and incubate with shaking at 37 °C for 3~4 h until the bacteria reach the logarithmic growth phase (OD600 $\approx$ 0.5), and place the bacterial liquid on ice for 30 min.
3. Centrifuge at 3000 r/min at 4 °C for 10 min to collect the bacteria, add 20 mL of pre-chilled 0.1 mol/L $CaCl_2$ solution to gently suspend the bacteria, and let stand on ice for 15 min.

4. Centrifuge at 3000 r/min at 4 °C for 10 min, discard the supernatant.
5. Add 2 mL of pre-chilled 0.1 mol/L $CaCl_2$ solution containing 15% glycerol, resuspend the cells and distribute them in 1.5 mL centrifuge tubes, 100 μL per tube.
6. Store at −80 °C.

3 Methods

Carry out all procedures at 4 °C unless otherwise specified.

3.1 Obtaining the dsRNA Sequence Fragment of the Target Gene

1. The GC content of the dsRNA sequence is more appropriate between 30% and 52%.
2. The length of the dsRNA sequence can be 200~4000 bp, and 300~800 bp is better.
3. Ensure that the selected dsRNA is highly specific in the target gene.
4. Fully consider the secondary structure of the sequence to avoid forming a stable hairpin structure.
5. Avoid consecutive single bases (e.g., AAAA) and nuclease cleavage sites (*see* **Notes 9** and **10**).
6. Avoid repetitive sequences (e.g., repetitive sequences, satellite DNA, transposons, etc.).
7. Generally select 50–100 nucleotide positions downstream of the start codon of the gene to start designing dsRNA.
8. The 3′ end sequence is preferred to the 5′ end sequence.
9. Fully compare the sequences of non-target species (mammals, natural enemy insects, humans, etc.) to avoid off-targets.
10. When designing dsRNA primers, add a T7 promoter (TAA TACGACTCACTATAGGG) at the 5′ end of the two primers (*see* **Notes 3** and **4**).

3.2 Construction of Recombinant Plasmids

3.2.1 L4440 Plasmid Extraction

1. Host cells containing L4440 were selected from the plate medium and inoculated into the liquid medium containing antibiotics for overnight culture at 37 °C (*see* **Note 8**).
2. Pour 5 mL bacterial solution into 5 mL centrifuge tube for 10,000 g, centrifuge for 1 min, discard supernatant.
3. Add 250 μL Solution I (add RNase A1 before use, refrigerate at 4 °C) to fully suspend bacterial precipitation (be careful not to retain small bacteria, and use Vortex and other oscillators to fully suspend bacteria).
4. Add 250 μL Solution II (put it into the incubator at 37 °C in advance to preheat) and gently turn it up and down and mix it

5~6 times to fully crack the bacteria and form a transparent solution (this step should not exceed 5 min).

5. Add 350 μL Solution III, gently turn up and down and mix for 5~6 times until a compact agglutination is formed.
6. Pour the obtained solution into 1.5 mL EP tube and centrifuge 13,000 g for 10 min.
7. Pour the centrifugal liquid into the column, 10,000 g, centrifuge for 1 min, discard the liquid.
8. Add 500 μL Buffer HB, 10,000 g, centrifuge for 1min, and discard the liquid.
9. Add 700 μL DNA wash buffer (add an equal volume of 100% ethanol in advance), centrifuge 10,000 g for 1 min, and discard the liquid (repeat once).
10. Centrifuge the empty tube at 13,000 g for 2 min after discarding the waste liquid.
11. Add 60 μL ddH_2O (preheated in advance in a 60 °C water bath) and shake at 60 °C for 2 min; 13,000 g, centrifuge for 1 min to obtain liquid.
12. Agarose electrophoresis test, confirm correct after −80 save, set aside.

3.2.2 Plasmid Ligation and Transformation

1. L4440 plasmid and the target gene fragment were simultaneously subjected to double digestion at 37 °C for 1 h, and the digestion products were recovered respectively (*see* **Notes 5–7** and **9**).
2. T4 ligase was used to connect the recovered L4440 vector to the target gene fragment at room temperature at a ratio of 1:3 for 10 min.
3. HT115 (DE3) competent cells were transformed, 100 μL competent cell suspension was taken from the −80 °C refrigerator (only 50 μL in the case of high efficiency), and thawed on ice. After thawing, 1~5 μL ligation product solution was added to the competent cell suspension, and it was gently mixed and placed on ice for 30 min.
4. Incubate in 42 °C water bath for 90 s, and quickly put on ice to cool for 3~5 min. Add 1 mL LB liquid medium (without antibiotics) to the tube, mix and shake at 37 °C for incubation for 1 h, so that the bacteria can resume growth. After shaking, take 100 μL coated on the screening plate containing antibiotics, and place it in the face for 30 min. After the bacterial solution was completely absorbed by the medium, the culture dish was inverted at 37 °C for 16~24 h.
5. Individual colonies were selected and cultured in liquid medium containing antibiotics. After the bacterial broth was

tested positive by PCR, the recombinant plasmid was extracted and identified by double enzyme digestion.

3.3 dsRNA-Induced Expression

1. Clone bacteria solution identified as positive in the previous step was inoculated into 100 mL LB medium at a volume ratio of 1:100, and incubated at 37 °C, 200 rpm for about 2 h.
2. When the OD600 of the culture broth was between 0.3 and 0.6, IPTG was added to a final concentration of 0.4 mmol/L, and continued to incubate for 4 h. Collect the culture broth for dsRNA isolation (*see* **Notes 13** and **14**).

3.4 dsRNA Extraction (see Notes 12 and 15).

3.4.1 Phenol Extraction by Chloroform

1. 50 mL of the induced expression bacterial solution was centrifuged at 7500 g for 5 min at 4 °C and precipitated with 100:1 volume of STE buffer for vortex oscillation.
2. Add 1:1 volume phenol: chloroform (25:24) mixture, vortex shake for 100 s, and centrifuge at 13,000 g for 15 min at 4 °C.
3. Add 2:1 volume isopropyl alcohol to the supernatant, leave at room temperature for 10 min and centrifuge at 4 °C for 10 min at 13,000 g.
4. Add 1:1 volume of 75% ethanol to the precipitate and centrifuge at 4 °C for 5 min at 7500 g.
5. Repeat **step 4** once or twice.
6. Precipitate for 5~10 min at room temperature, add 50 μL DEPC treatment water.
7. Add 10:1 volume diluted RNase A and DNase I, incubate at 37 °C for 15 min for digestion. Store at −80 °C after electrophoresis detection.

3.4.2 One-Step Extraction (75%Alcohol Precipitation Method)

1. Take 50 mL of the induced bacterial solution, centrifuge at 7500 g for 2 min at 4 °C and discard the supernatant.
2. Add 25:1 volume of 75% ethanol (PBS), blow and mix well, and leave at room temperature for 5 min.
3. Centrifuge at 6000 g for 5 min at 4 °C, and discard the supernatant.
4. The precipitate was suspended with a 4:1 volume of 150 mmol/L NaCl solution and stood at room temperature for more than 60 min.
5. Centrifuge 13,000 g at 4 °C for 10 min to obtain the supernatant.
6. Add 10:1 volume diluted RNase A and DNase I for digestion, and incubate at 37 °C for 15 min.
7. Store at −80 °C after electrophoretic detection.

3.4.3 TRIzol Method

1. Take 50 mL of the induced bacterial solution and centrifuge at 4 °C for 5 min at 12,000 r/min and discard the supernatant.
2. Precipitate and add 50:1 volume of TRIzol, shake vigorously, and then leave at room temperature for 5 min.
3. Add 5:1 volume of chloroform, and leave for 10 min at room temperature and 15 min at 12,000 g at 4 °C.
4. Add 2:1 volume isopropyl alcohol to the supernatant, reverse 5–6 times, and repeat once.
5. Precipitation with 1:1 volume 75% ethanol at 4 °C for 7500 g for 5 min and repeat once.
6. Precipitate at room temperature for 5~10 min, add 50 μL DEPC treatment water.
7. Add 10:1 volume diluted RNase A and DNase I for digestion, incubate at 37 °C for 15 min, and store at −80 °C after electrophoresis detection.

4 Notes

1. Adjusting the pH of the medium to 7.4 using NaOH as a rule of thumb (this pH is more suitable for the growth of *E. coli*).
2. During the cooling process of solid medium, the UV lamp is irradiated for 10~15 min to ensure a sterile environment.
3. A carrier system consisting of a constitutive promoter and a tandem T7 promoter to maximize dsRNA synthesis.
4. When the two promoters run simultaneously and are accumulated by dsRNA, the metabolic cost of bacteria is the highest.
5. The distance between the two cleavage sites on the selected plasmid should not be less than too small (>10 bp), otherwise it will affect the recognition of the cleavage points by restriction enzymes, which is not conducive to the double cleavage of empty vectors.
6. The selection of restriction sites for the enzyme should be taken into account when the general target gene is used for clonal expression.
7. The target gene fragment does not contain the selected cleavage site (otherwise the target gene will be cut off when the positive recombinants are identified by double cleavage).
8. All vectors (eukaryotic, prokaryotic, yeast, insect) for subsequent application of the experiment should contain the selected cleavage site whenever possible and ensure that the insertion on the vector is in the correct direction (directed cloning).
9. Two enzyme cleavage points are separated by at least 3 bases. The two enzyme cutting points should not be the same tail

enzyme (cut-off residues do not complement each other), otherwise the effect is equivalent to single enzyme cutting.

10. It is best to use enzymes with high enzyme-cutting efficiency, and use double enzymes to cut enzymes with a common buffer.
11. The host bacteria culture is not easy to be excessive, and excessive bacteria content will lead to insufficient melting bacteria, which will affect the purity of the plasmid during purification.
12. After dsRNA is extracted, it should be digested with RNase A and DNase I.
13. IPTG can be replaced with cheaper and non-toxic natural materials (lactose, skim milk, etc.).
14. Avoid contamination by bacteria, fungi and phages during fermentation.
15. dsRNA extraction efficiency: Phenol extraction by chloroform > TRIzol method > One-step extraction.

References

1. Christiaens O, Niu J, Nji Tizi Taning C (2020) RNAi in insects: a revolution in fundamental research and pest control applications. Insects 11
2. Delgado-Martin J, Velasco L (2021) An efficient dsRNA constitutive expression system in Escherichia coli. Appl Microbiol Biotechnol 105:6381–6393
3. Dvorak P, Chrast L, Nikel PI, Fedr R, Soucek K, Sedlackova M, Chaloupkova R, de Lorenzo V, Prokop Z, Damborsky J (2015) Exacerbation of substrate toxicity by IPTG in Escherichia coli BL21(DE3) carrying a synthetic metabolic pathway. Microb Cell Factories 14:201
4. Guan R, Chu D, Han X, Miao X, Li H (2021) Advances in the development of microbial double-stranded RNA production systems for application of RNA interference in agricultural pest control. Front Bioeng Biotechnol 9:753790
5. Guan R, Li H, Miao X (2022) Commercialization status and existing problems of RNA biopesticides. Sci Agric Sin 55:2949–2960
6. He L, Huang Y, Tang X (2022) RNAi-based pest control: production, application and the fate of dsRNA. Front Bioeng Biotechnol 10:1080576
7. Khani MH, Bagheri M (2020) Skimmed milk as an alternative for IPTG in induction of recombinant protein expression. Protein Expr Purif 170
8. Ma ZZ, Zhou H, Wei YL, Yan S, Shen J (2020) A novel plasmid-Escherichia coli system produces large batch dsRNAs for insect gene silencing. Pest Manag Sci 76:2505–2512
9. Nwokeoji AO, Nwokeoji EA, Chou T, Togola A (2022) A novel sustainable platform for scaled manufacturing of double-stranded RNA biopesticides. Bioresour Bioprocess 9
10. Papic L, Rivas J, Toledo S, Romero J (2018) Double-stranded RNA production and the kinetics of recombinant Escherichia coli HT115 in fed-batch culture. Biotechnol Rep (Amst) 20:e00292
11. Taning CN, Arpaia S, Christiaens O, Dietz-Pfeilstetter A, Jones H, Mezzetti B, Sabbadini S, Sorteberg HG, Sweet J, Ventura V et al (2020) RNA-based biocontrol compounds: current status and perspectives to reach the market. Pest Manag Sci 76:841–845
12. Vogel E, Santos D, Mingels L, Verdonckt TW, Broeck JV (2018) RNA interference in insects: protecting beneficials and controlling pests. Front Physiol 9:1912
13. Yin G, Sun Z, Liu N, Zhang L, Song Y, Zhu C, Wen F (2009) Production of double-stranded RNA for interference with TMV infection utilizing a bacterial prokaryotic expression system. Appl Microbiol Biotechnol 84:323–333

Chapter 12

Transiently Induce RNA Silencing in Plants Using a Tobacco Necrosis Virus A (TNV-A)-Based dsRNA Production System

Yuanming Zhang, Mengzhu Chai, Xiaofei Cheng, and Kai Xu

Abstract

Transgenic expression of hairpin RNA or artificial microRNA is widely used for genetic studies in plant science. However, induction of RNA silencing by transgenic method may have a problem when studying essential genes. Here, we provide an *in planta* transient double-stranded RNA (dsRNA) producing system using a tobacco necrosis virus A (TNV-A)-based replicon for efficiently inducing RNA silencing in plants. In this system, the target sequence is placed between the cauliflower mosaic virus 35S promoter and the 3′-terminal part of viral genomic RNA, while the C-terminal part of TNV-A RNA-dependent RNA polymerase (p82C) is expressed by a different promoter. The endogenous RNA polymerase-synthesized target sequence is recruited by p82C to produce dsRNA to induce RNA silencing.

Key words Tobacco necrosis virus, dsRNA, Replicon, RNA-dependent RNA polymerase, Hairpin structure

1 Introduction

RNA silencing, also known as RNA interference (RNAi), is an evolutionarily conserved sequence-specific gene regulatory and antiviral mechanism [1, 2]. In RNA silencing, double-stranded RNAs (dsRNAs), which can either be synthesized by viral or endogenous RNA-dependent RNA polymerase, formed by renaturation of two complementary RNA molecules, or folded in an RNA molecule containing internal inverted-repeated sequence [3, 4], is recognized and processed by Dicer or Dicer-like (DCL) riboendonucleases, members of the ribonuclease III family, into 21–24 nucleotide (nt) small RNA duplexes [5]. These small RNA duplexes are then methylated by HEN1, and one strand of these small RNA duplexes is integrated into Argonate (AGO) proteins, which func-

Yuanming Zhang and Mengzhu Chai contributed equally with all other contributors.

Xiaofei Cheng and Guanwei Wu (eds.), *Double-Stranded RNA: Methods and Protocols*, Methods in Molecular Biology, vol. 2771, https://doi.org/10.1007/978-1-0716-3702-9_12,

tion through base pairing, e.g., cleavage of target mRNA, inhibit the transcription or translation of target gene, or directing the methylation of target DNA [6–8].

In plants, RNA silencing can be triggered by stable or transient expression of hairpin RNA or artificial microRNA. Stable expression of hairpin RNA or artificial microRNA requires time-consuming transformation procedure, and sometimes is an impossible mission in some plants. Transient expression of hairpin RNA is an effective meaning for inducing the posttranscriptional gene silencing (PTGS); however, it often has only a limited silencing effect [9]. Given the factor that viruses are targeted by RNA silencing, viruses are also ideal agents for inducing RNA silencing. More than a dozen plant viruses have been developed into vectors for inducing RNA silencing on varied plant species [10]. However, viruses always encode viral suppressor of RNA silencing and other components that may reduce the efficiency of RNA silencing. Moreover, the movement and systemic infection of full virus may cause the necrosis and even death of the plant, which could cause the failure of the experiment. Viral replicon or a deconstructed viral vector is an alternative for this purpose [11]. In this chapter, we describe the modification of an alphanecrovirus of the family *Tombusviridae*, namely tobacco necrosis virus A (TNV-A) into a deconstructed viral replicon for transient production of dsRNA in plant cells to induce efficient RNA silencing.

2 Materials

2.1 Materials and Reagents

1. The empty plasmid pCB301-314-pTNV(+)-RZ containing multicloning sites (MCS), TNV-A positive-strand initiating promoter sequence, and Hepatitis delta virus antigenomic ribozyme (Rz) under the Cauliflower mosaic virus (CaMV) 35S promoter (*see* **Note 1**).
2. The plasmid pGD-2 × 35S-TNV-p82C expressing the C-terminal of TNV-A-encoded p82 (p82C) under double CaMV 35S promoter (*see* **Note 1**).
3. The hairpin control plasmid pGD-35S-chsA-NbPDS-hairpin [12].
4. Chemically competent cells of *Escherichia coli* strains DH5α.
5. Electrocompetent cells of *Agrobacterium tumefaciens* strain GV3101 (*see* **Note 2**).
6. Primer premier 5.0 software.
7. *Nicotiana benthamiana* seeds.
8. Plant total RNA kit (e.g., Merck Sigma-Aldrich Spectrum™ Plant Total RNA Kit, Cat # STRN10).

9. cDNA Synthesis kit (e.g., Thermo Fisher Scientific PureLink Plant RNA kit, Cat # 12322012).
10. Kit for DNA fragment recovery from gel (e.g., Qiagen QIAquick Gel Extraction Kit, Cat # 28704).
11. High fidelity DNA polymerase (e.g., Thermo Fisher Scientific Phusion High fidelity DNA polymerase 2 U/μL, Cat # F530S).
12. Taq DNA polymerase (e.g., Thermo Fisher Taq DNA polymerase, recombinant, Cat # 10342053).
13. Plasmid purification kit (e.g., Qiagen QIAprep Spin Miniprep Kit, Cat # 27104).
14. Quantitative PCR kit (e.g., Vazyme ChamQ Blue Universal SYBR qPCR Master Mix, Cat # Q312-02).
15. Kanamycin (Kana; Shanghai Sangon, Catalog # A00408).
16. Rifampicin (Rif; Shanghai Sangon, Catalog # A600812).
17. Gentamycin sulfate (Gent; Shanghai Sangon, Catalog # B540724).
18. LB medium: weight 10 g tryptone, 10 g sodium chloride (NaCl), and 5 g yeast extract, added deionized water to 1 L, autoclave at 121 °C for 15–20 min. Add 1% agar for making LB plate.
19. 1.0 M acetosyringone stock solution: Dissolve 1.962 g of acetosyringone in 9 mL dimethylsulfoxide (DMSO), and top up to 10 mL with DMSO; Filter sterilize and store at −20 °C from light (*see* **Note 3**).
20. Infiltration buffer (10 mM $MgCl_2$, 150 μM acetosyringone, and 5 mM MES, pH = 5.6): Take 100 μl 0.5 M MgCl and 100 μL 1.0 M 2-(N-morpholino)ethanesulfonic acid (MES; pH 5.6) in a 10 mL measuring cylinder, top up to 10 mL with autoclaved distilled water; add 1.5 μL 1.0 M acetosyringone stock solution before use.
21. Restriction enzymes *Stu*I and *Sal*I.
22. T4 DNA ligase.

2.2 Equipment

1. Pots and soil mixture (e.g., SunGro Sunshine mix #1).
2. Plant growth chamber.
3. Thermal cycler (PCR machine; e.g., LongGene T30 Thermal Cycler).
4. Water bath.
5. Measuring cylinder.
6. Pipette.
7. Benchtop Centrifuge.

8. Shaking incubator.
9. Lab Incubator.
10. Autoclave.
11. Gel electrophoresis device.
12. Electroporator and cuvettes (e.g., Bio-Rad MicroPulser Electroporator and 0.1 cm Electroporation Cuvette).
13. Quantitative PCR machine (e.g., Bio-Rad CFX96 touch).
14. Microvolume spectrophotometer (e.g., Thermo Fisher Scientific NanoDrop 8000 Spectrophotometer).
15. Visible Spectrophotometer (e.g., Eppendorf BioPhotometer UV/VIS).

3 Methods

3.1 Plasmid Construction

1. Download the target sequence (e.g., *N. benthamiana PHYTOENE DESATURASE* gene) from the GenBank database and design primers with the Primer premier 5.0 (*see* **Note 4**).
2. Extract the total RNA from *N. benthamiana* leaves according to the user manual of the RNA isolation kit.
3. Synthesize complementary DNA (cDNA) using Oligo-dT_{18} primer according to the product protocol.
4. Amplify the *PDS* fragment and with high-fidelity DNA polymerase with the primers from **step 1**.
5. Analyze the PCR product in 1% agarose gel by electrophoresis.
6. Excise the amplicon from the agarose gel with a clean scalpel and purify the target DNA using a DNA gel extraction kit.
7. Digest the PCR product and pCB301-314-pTNV(+)-RZ plasmid with *Stu*I and *Sal*I.
8. Perform the ligation reaction to insert the fragment of target gene into pCB301-314-pTNV(+)-RZ (e.g., pCB301-*PDSi*-pTNV(+)-RZ) (*see* **Note 5**).
9. Perform *E. coli* heat shock transformation.
10. Select the correct colony by PCR with Taq DNA polymerase.
11. Culture correct colony, extract plasmid, and send out for sequencing to confirm the plasmid.

3.2 Plant Inoculation

1. Transfer pGD-2 × 35S-TNV-p82C and pCB301-*PDSi*-pTNV(+)-RZ into *Agrobacterium* strain GV3101 by electroporation (*see* **Note 6**).
2. Plate the bacteria mixture on LB plate containing 50 mg/L Kana, 20 mg/L Gent, and 10 mg/L Rif and then culture the plate at 28 °C for 2–3 days.

3. Select the correct colonies by colony PCR.
4. Culture the correct bacteria in liquid LB medium containing 50 mg/L Kana, 20 mg/L Gent, and 10 mg/L Rif at 28 °C overnight in a shaking incubator.
5. Centrifuge at 6000 rpm to collect the bacteria at room temperature.
6. Wash the bacterial pellets twice by infiltration buffer.
7. Resuspend the bacteria and adjust the O.D. 600 to 1.0 with infiltration buffer (*see* **Note 7**).
8. Place the bacterial solution at room temperature for at least 2 h.
9. Measure equal volumes to the bacterial solution harboring pGD-2 × 35S-TNV-p82C and pCB301-*PDSi*-pTNV(+)-RZ, mix well by turning up and side down several times, and infiltrated into leaves of four-week-old seedlings of *N. benthamiana* (*see* **Note 8**).

3.3 RNA Silencing Evaluation

1. Collect the agroinfiltrated leaf area at about 3 days post-inoculation (dpi) (*see* **Note 9**).
2. Extract the total RNA with a total plant RNA isolation kit according to the supplied protocol.
3. Synthesize the first strand of cDNA using oligo-dT18 primer.
4. Qualify the concentration of cDNA with NanoDrop2000.
5. Set up quantitative PCR with an equal amount of cDNA.

3.4 Results

1. The 67–502 bp of *N. benthamiana PHYTOENE DESATURASE* (*PDS*) gene (Genbank accession no. DQ469932.1) was chosen as the example in this protocol. The endonuclease restriction site *Stu*I and *Sal*I were added to the primers *NbPDSi-F* (5′-GAAAGGCCTAGCTCGAGGTCTTCGTTGGGAAC-3′) and *NbPDSi-R* (5′-ACGCGTCGACAGAATATGTGCAACCCAGTCTCGTACC-3′), respectively (*see* **Note 10**).
2. The following pair of primers were used for determining the relative transcript level of *NbPDS* in quantitative PCR (qPCR): NbPDS-rtF (5′-AGGGTGACAGATGAGGTGTT-3′) and NbPDS-rtR (5′-TTCCACAATCGGCATGCAAA-3′). qPCR was performed with the *NbTubulin* as the internal control, which was determined using the following two primers: NbTubulin-rtF (5′-GGTTGGACATTACCGTGATACC-3′) and NbTubulin-rtR (5′-ACTTCATGTCAAGCTCGATTTCT-3′) (*see* **Note 11**).
3. The qPCR was performed using the ChamQ SYBR Color qPCR Master Mix and 500 ng cDNA. Gene expression levels were calculated using the $2^{-\Delta\Delta Ct}$ algorithm. Based on the

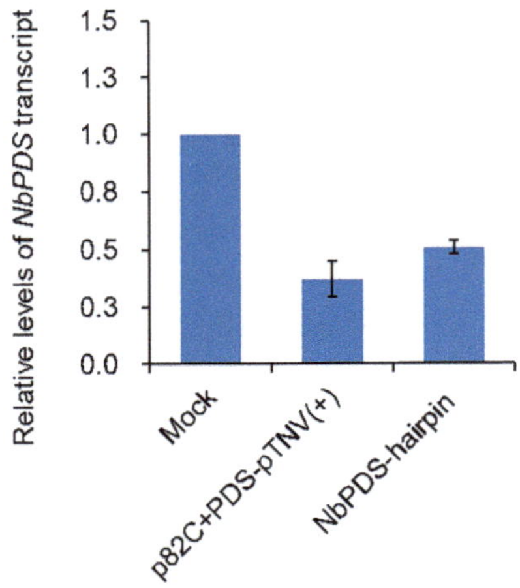

Fig. 1 Bar chart showing the relative transcript levels of *NbPDS* in the leaf area infiltrated with different combination of Agrobacteria. Data are means of three biological repeats. The *NbTubulin* was used as the internal control

qPCR results, it is clear that transcript level of *NbPDS* was reduced by more than 60% in the leaf area expressing pGD-2 × 35S-TNV-p82C and pCB301-*PDSi*-pTNV(+)-RZ, while only about 50% percentage was silenced by hairpin structure (Fig. 1).

4 Notes

1. The plasmids pCB301-314-pTNV(+)-RZ and pGD-2 × 35S-TNV-p82C were described in [12].
2. Do not use the one harboring suppressor of RNA silencing, e.g., tomato bushy stunt virus (TBSV)-encoded p19.
3. Ethanol can be used to dissolve acetosyringone if no DMSO is available.
4. Do not forget to add the restriction enzyme site *Stu*I at the 5′-end of forward primer and *Sal*I at the 5′-end of reverse primer. Do not forget to add the homologous arm at both ends of primers if recombinant cloning is used.
5. Always perform the ligation reaction at 16 °C.
6. The plasmid pGD-35S-chsA-NbPDS-hairpin that expresses a hairpin structure of *NbPDS* was included as a hairpin RNA control [12].
7. This step can increase the transformation efficiency of Agrobacteria.
8. For the control experiment, Agrobacteria harboring pGD-35S-chsA-NbPDS-hairpin were mixed with those harboring pGD-2 × 35S-TNV-p82C to ensure the same loading of Agrobacteria.

9. The infiltrated leaf area may display no obvious or slight bleaching phenotype depending on the experimental conditions and statutes of the *N. benthamiana* plants.
10. The fragment should have a length above 250 bp but shorter than 1000 bp for the best silencing efficiency.
11. The primers used for qPCR should be located outside the region used for silencing.

References

1. Plasterk RHA (2002) RNA silencing: the genome's immune system. Science 296(5571): 1263–1265
2. Baulcombe D (2005) RNA silencing. Trends Biochem Sci 30(6):290–293
3. Maida Y, Masutomi K (2011) RNA-dependent RNA polymerases in RNA silencing. Biol Chem 392(4):299–304f
4. Rosok O, Sioud M (2004) Systematic identification of sense-antisense transcripts in mammalian cells. Nat Biotechnol 22(1):104–108
5. Schauer SE, Jacobsen SE, Meinke DW, Ray A (2002) DICER-LIKE1: blind men and elephants in Arabidopsis development. Trends Plant Sci 7(11):487–491
6. Meister G (2013) Argonaute proteins: functional insights and emerging roles. Nat Rev Genet 14(7):447–459
7. Hutvagner G, Simard MJ (2008) Argonaute proteins: key players in RNA silencing. Nat Rev Mol Cell Biol 9(1):22–32
8. Yang Z, Ebright YW, Yu B, Chen X (2006) HEN1 recognizes 21–24 nt small RNA duplexes and deposits a methyl group onto the 2′ OH of the 3′ terminal nucleotide. Nucleic Acids Res 34(2):667–675
9. Wesley SV et al (2001) Construct design for efficient, effective and high-throughput gene silencing in plants. Plant J 27(6):581–590
10. Robertson D (2004) VIGS vectors for gene silencing: many targets, many tools. Annu Rev Plant Biol 55:495–519
11. Gleba Y, Marillonnet S, Klimyuk V (2004) Engineering viral expression vectors for plants: the 'full virus' and the 'deconstructed virus' strategies. Curr Opin Plant Biol 7(2):182–188
12. Zhang W et al (2021) Development of a viral RdRp-assisted gene silencing system and its application in the identification of host factors of plant (+)RNA virus. Front Microbiol 12: 682921

Chapter 13

Co-immunoprecipitation-Based Isolation of Double-Stranded RNA-Associated Protein Complexes in *Nicotiana benthamiana*

Zhaoxing Jia, Jianping Chen, Fei Yan, and Guanwei Wu

Abstract

Double-stranded RNA (dsRNA) is associated with most viral infections, and is generated in host cells during viral replication. Viral RNA replication occurs within the viral factories called the viral replication complexes (VRCs). In addition to viral genome, viral-derived dsRNA and replicase, the VRCs composition remains largely unexplored. The dsRNA binding domain of the B2 protein from Flock house virus has been reported to be used for detecting viral-derived long dsRNA in plants efficiently. *Nicotiana benthamiana* is widely used as a model plant for plant-microbe interactions owing to its susceptibility to diverse plant diseases, especially viral diseases. Here, we describe the use of *Nicotiana benthamiana* stably expressing GFP-tagged dsRNA binding protein (B2: GFP) to pull down dsRNA and associated host and viral proteins from turnip mosaic virus-infected plants. The obtained protein complexes are compatible with functional assays, Western blotting, and mass spectrometry. This system provides a valuable and robust tool to study VRC proteome in *N. benthamiana* upon plant viral infections.

Key words Double-stranded RNA, *Nicotiana benthamiana*, Protein-complex immunoprecipitation, GFP antibody, dsRNA binding protein B2, Turnip mosaic virus

1 Introduction

As obligate pathogens, plant viruses depend on host factors and their related metabolism machinery to establish successful infection in plants [1]. Generally, the infection process of plant virus in host cells can be divided into several stages, including virion uncoating, translation, replication, viral particle assembly, cell-to-cell and long-distance movement [2]. Viral replication is one of the most critical steps to establish virus infection in host cells [3]. Viral replication is carried out by viral-encoded RNA-dependent RNA polymerase (RdRp), which is essential for transcription and replication of their genome since host cells lack equivalent enzymes. Viral RdRP firstly replicates the virus (+) genome into (−) single-stranded

Xiaofei Cheng and Guanwei Wu (eds.), *Double-Stranded RNA: Methods and Protocols*, Methods in Molecular Biology, vol. 2771, https://doi.org/10.1007/978-1-0716-3702-9_13,

(ss) RNA, and then uses it as a template to produce offspring (+) ssRNA. In fact, the essence of viral RNA replication process is to produce long double-stranded RNA (dsRNA) intermediates from viral RdRp. Replication of positive-strand RNA viruses often occurs in organelle membranes [4], such as endoplasmic reticulum (ER) [5], chloroplast [6], mitochondrion [7], peroxisome [8], or vacuole. In the process of virus replication, viruses use the hosts' membrane system to establish its own replication sites, forming a relatively safe compartment to protect the virus genome from the interference of the host's antiviral mechanism [9]. These relatively safe compartments formed by the membrane system from the host are called viral replication complexes (VRCs). The rearrangement of host membrane structure to form VRCs is a very complicated and meticulous process, which needs to be completed with the participation of many host factors [3, 4, 10]. Although the simplest and smallest VRCs are composed of single-stranded and double-stranded viral RNA and replicase, their exact composition changes with the difference between virus and host species, which has not been explored to a large extent. Accumulated studies have proved that some host proteins are components of these VRCs and play a positive (pro-viral) or negative (antiviral) role in replication [3, 11, 12]. Therefore, identifying and mining the host factors involved in the formation of VRCs is of great significance for further understanding of viral replication, and will help to develop anti-virus strategies based on mechanisms.

Recently, researchers report that the GFP-tagged dsRNA-binding protein domain of FHV protein B2 can be used to visualize VRCs in *N. benthamiana* [13], and also exploited to pull-down dsRNA-associated proteome from Arabidopsis plants upon infected with tobacco rattle virus [14]. Protein-complex immunoprecipitation (co-IP) has been extensively used to study protein-protein interactions. This method relies on the transient expression of recombinant epitope-tagged proteins in *Nicotiana benthamiana*, followed by protein extraction, co-IP, SDS-PAGE, and mass spectrometry. Here, we present a detailed protocol for the isolation of dsRNA-associated host and viral protein complexes in *Nicotiana benthamiana* B2:GFP plants upon turnip mosaic virus infection.

2 Materials

2.1 Plant Material and Growth Conditions

Nicotiana benthamiana plants stably expressing B2:GFP or GFP were grown in a growth-controlled chamber at 23 to 24 °C under fluorescent lights at 16 h/8 h light/dark photoperiod. Healthy plants showing green fluorescences under hand-held UV-lamp with 6–8 leaves should be considered for agro-infiltration (*see* **Note 1**).

2.2 Preparation of Agrobacterium Culture

Agrobacterium tumefaciens strain GV3101 is cultured on solid or liquid Luria–Bertani (LB) medium: 1% (W/V) Bacto Tryptone, 0.5% (W/V) yeast extract, 1% (W/V) NaCl, and adjust pH to 7.00 with NaOH. Autoclave at 121 °C for 15 min.

2.3 Agrobacterium Infiltration Buffer

10 mM $MgCl_2$, 10 mM MES (2-(N-morpholino) ethanesulfonic acid) (pH 5.6), 200 μM acetosyringone (Millipore Sigma) (*see* **Note 2**).

2.4 Protein Extraction Buffer and Co-immunoprecipitation Reagents

1. GTEN buffer: 10% glycerol, 25 mM Tris pH 7.5, 1 mM EDTA, 150 mM NaCl.
2. Extraction buffer: GTEN, 2% w/v PVPP, 10 mM DTT, 1× protease inhibitor cocktail (Sigma), 0.1% IGEPAL® CA-630 (Sigma).
3. ChromoTek GFP-Trap® Magnetic Agarose (proteintech).
4. Immunoprecipitation (IP) buffer: GTEN, 0.1% IGEPAL® CA-630 (Sigma), 1× protease inhibitor cocktail (Sigma) (*see* **Note 3**).

2.5 SDS-Polyacrylamide Gel Electrophoresis

1. Separating buffer: 1.5 M Tris–HCl, pH 8.8.
2. Stacking buffer: 1 M Tris–HCl, pH 6.8.
3. 10% (w/v) SDS.
4. 10% (w/v) Ammonium persulfate (APS, Bio-Rad). The solution can be stored at 4 °C.
5. Thirty percent acrylamide/bis solution (29:1) (Bio-Rad, Hercules, CA).
6. N, N, N′, N′-Tetramethylethylenediamine (TEMED, Bio-Rad).
7. Running buffer (10×): Dissolve 30.3 g Tris-base (MW121.14), 151.1 g glycine (MW75.07), 10 g SDS in 1 L sterile water.
8. Blue Plus® II Protein Marker (14–120 kDa) (TransGen Biotech).
9. Coomassie Blue Solution:0.1% (MW) Coomassie Bright Blue R-250, 25%(V/V) isopropyl alcohol, 10% Acetate acid gracial.
10. SDS-PAGE loading buffer (5×): 0.25 M Tris-Cl (pH 6.8), 10% (w/v) SDS, 50% (v/v) glycerol, 5% β-Mercaptoethanol, 0.5% bromophenol blue.

2.6 Western Blotting

1. Transfer buffer: 25 mM Tris-base, 192 mM glycine, 0.1% (w/v) SDS, 20% (v/v) Methanol, pH 8.3.
2. Immobilon®-P PVDF Membrane 0.45 μm.
3. Mini Trans-Blot Cell (Bio-Rad).

4. Tris-buffered saline with Tween (TBS-T) (10×): 1.37 M NaCl, 250 mM Tris-Cl, pH 7.5, 1% (v/v) Tween 20.
5. Blocking solution: 5% (w/v) non-fat milk powder in 1× TBS.
6. Primary antibody: Anti-GFP Mouse Monoclonal Antibody (TransGen Biotech).
7. Secondary antibody: ProteinFind® Goat Anti-Mouse IgG (H + L), HRP Conjugate (TransGen Biotech).

3 Methods

3.1 Agrobacterium Infiltration

1. Wild-type TuMV infectious clone PCB301-TuMV was transformed into Agrobacterium component cells and set 10 ml overnight liquid cultures of transformed Agrobacterium (*see* **Note 4**).
2. Pellet down the Agrobacterium cells by centrifugation at 4000 rpm for 5 min at room temperature, and wash the precipitates with 5 mL infiltration buffer. Repeat the steps twice.
3. Measure the OD_{600} of the cells and dilute to the value of 0.3–0.5.
4. Infiltrate the prepared Agrobacterium cells to the abaxial side of the leaf using a needless infiltration syringe. After infiltration, mark the zone of infiltration areas using a marker.

3.2 Sample Preparation and Protein Extraction

1. The leaves of plants (about 1.0 g) were sampled, and placed in a sterile EP tube with sterile steel balls, and then quickly put into liquid nitrogen for freezing storage.
2. Put the EP tube on a grinder and grind it for 90 s at 30–35-Hz until the leaf tissue is powdery.
3. Add GTEN extraction buffer to the EP tubes (according to 1 g leaf/2 mL GTEN extraction buffer).
4. Incubate EP tubes at 4 °C for 0.5–1 h (*see* **Note 5**).
5. Centrifuge at 4000 rpm for 10 min at 4 °C, and then transfer the supernatant to a new EP Tube and centrifuge at 12,000 rpm for 20–30 min at 4 °C. The resulting supernatant was used to proceed further.

3.3 Co-Immunoprecipitation

1. Resuspend 20 μL/reaction volume of the GFP magnetic beads with 500 μL IP buffer, and separate the magnetic beads from the IP buffer solution using a magnetic rack and remove the supernatant. Repeat the wash step 3 times. Maintain the magnetic beads on ice until the next step.

2. Resuspend the prepared magnetic beads with extracted total protein supernatant and mix by vortexing briefly. Incubate the mixture on a rotating mixture for 2 h at 4 °C. Place the magnetic beads-protein complex containing EP tubes on a magnetic rack, and remove the supernatant.
3. Add 1 mL IP buffer and wash the magnetic beads for 5–8 times (*see* **Note 6**).
4. Add 1 × SDS-PAGE running buffer and heat the mixture in a dry bath for 15 min at 75 °C. The resulting samples are used for SDS-PAGE analysis.

3.4 SDS-Polyacrylamide Gel Electrophoresis and Western Blotting Analysis

1. Preparation of 12% SDS-PAGE protein vertical gel: firstly, prepare 12% separation gel solution, mix it evenly, then pour it into the gap of the glue-making glass plate, flatten the glue surface with anhydrous alcohol, after the gel solidifies, pour off the anhydrous alcohol and absorb the residual liquid with absorbent paper, then add the prepared 5% concentrated gel solution, insert a comb, and wait for the gel to solidify before use.
2. Suck 5–10 μL of protein sample into the glue hole of the protein gel, and at the same time, preview the protein Marker as a reference for electrophoresis at 80 V for 30 min. After the sample runs out of the concentrated gel, electrophoresis at 100 V for 1–2 h.
3. Place the gel in transfer buffer and assemble the transfer sandwich. Make sure no air bubbles are trapped in the sandwich. The blot should be on the cathode and the gel on the anode.
4. Place the cassette in the transfer tank and transfer at 100 V for 1.5–2 h in a cold room. Stain the blot with Ponceau S solution to check the transfer quality and then rinse off the stain with TBST.
5. Briefly rinse the blot in water and block in 5% BSA in TBST at room temperature for 30 min.
6. Incubate in the anti-GFP solution at room temperature for 1.5–2 h and wash the blot 3–4 times for 3–5 min with TBST.
7. Incubate in the HRP-conjugated secondary antibody solution for 1 h at room temperature and wash the blot 3–4 times for 3–5 min with TBST.
8. Apply the chemiluminescent substrate to the blot and capture the chemiluminescent signals using a CCD camera-based imager.

4 Notes

1. The age and health of the plant are extremely important for successful infection of TuMV. The use of flowering plants must be avoided. *N. benthamiana* plants expressing B2:GFP or GFP should be examined by hand-held UV lamp prior to viral infection.
2. All the reagents and buffers should be prepared fresh using sterile Milli-Q water and molecular biology-grade chemicals.
3. IP buffer is critical for successful co-IP experiment. A higher concentration of IGEPAL® CA-630 in the buffer may abolish the interaction between proteins. Therefore, it is recommended to optimize the concentration. The specified concentration of NaCl in the wash buffer is compatible with most applications; however, the concentration can be increased to reduce nonspecific proteins or decreased to prevent washing off desired proteins.
4. Freshly transformed Agrobacterium is recommended to use for better viral infection, and TuMV CP accumulation can be detected at 3 dpai (Fig.1b).
5. Majority of the steps in this protocol are suggested to be performed at 4 °C, the incubation times will vary and must be optimized.
6. Magnetic beads-protein complexes should be washed robustly to minimize non-specific complex binding. Clear bands of GFP and B2-GFP with expected sizes can be detected (Fig.1a).

Acknowledgments

We are indebted to Prof. Christophe Ritzenthaler (Université de Strasbourg, France) for the transgenic T2 dsRNA reporter *N. benthamiana* line B2-GFP. This research was funded by grants from the National Natural Science Foundation of China (3207016) to GW, the China Agriculture Research System of MOF and MARA (CARS-24-C-04) to FY, and the K. C. Wong Magna Fund in Ningbo University.

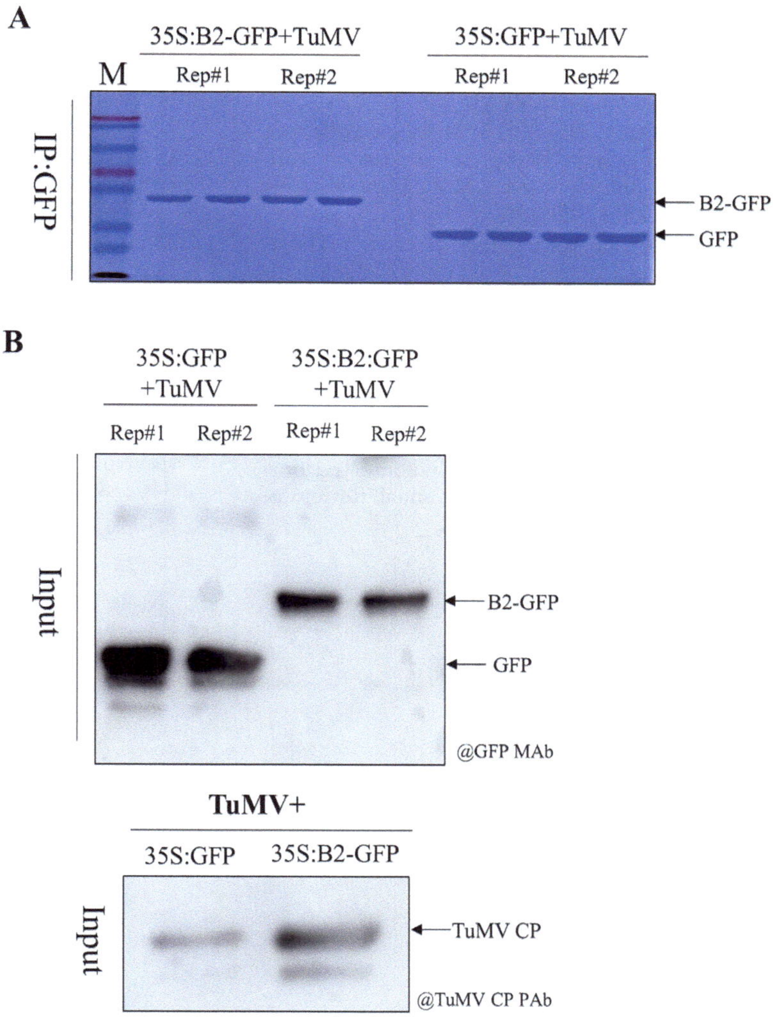

Fig. 1 Co-immunoprecipitation of double-stranded RNA-associated protein complexes. SDS-PAGE (**a**) and Western blotting (**b**) analysis of the co-IP products using GFP magnetic beads from *N. benthamiana* plants stably expressing B2: GFP or GFP upon TuMV infection

References

1. Wang A (2015) Dissecting the molecular network of virus-plant interactions: the complex roles of host factors. Annu Rev Phytopathol 53:45–66
2. Saxena P, Lomonossoff GP (2014) Virus infection cycle events coupled to RNA replication. Annu Rev Phytopathol 52:197–212
3. Nagy PD, Pogany J (2011) The dependence of viral RNA replication on co-opted host factors. Nat Rev Microbiol 10(2):137–149
4. Verchot J (2011) Wrapping membranes around plant virus infection. Curr Opin Virol 1(5):388–395
5. Cao X et al (2015) Morphogenesis of endoplasmic reticulum membrane-invaginated vesicles during beet black scorch virus infection: role of auxiliary replication protein and new implications of three-dimensional architecture. J Virol 89(12):6184–6195

6. Wang X et al (2021) Three-dimensional reconstruction and comparison of vacuolar membranes in response to viral infection. J Integr Plant Biol 63(2):353–364
7. Gomez-Aix C et al (2015) Melon necrotic spot virus replication occurs in association with altered mitochondria. Mol Plant-Microbe Interact 28(4):387–397
8. Fernandez de Castro I et al (2017) Three-dimensional imaging of the intracellular assembly of a functional viral RNA replicase complex. J Cell Sci 130(1):260–268
9. Laliberté J-F, Sanfaçon H (2010) Cellular remodeling during plant virus infection. Annu Rev Phytopathol 48(1):69–91
10. Nagy PD (2016) Tombusvirus-host interactions: co-opted evolutionarily conserved host factors take center court. Annual Review of Virology 3(1):491–515
11. Nagy PD (2020) Host protein chaperones, RNA helicases and the ubiquitin network highlight the arms race for resources between tombusviruses and their hosts. Adv Virus Res 107: 133–158
12. Li F, Wang A (2019) RNA-targeted antiviral immunity: more than just RNA silencing. Trends Microbiol 27(9):792–805
13. Monsion B et al (2018) Efficient detection of long dsRNA in vitro and in vivo using the dsRNA binding domain from FHV B2 protein. Front Plant Sci 9:70
14. Incarbone M et al (2021) Immunocapture of dsRNA-bound proteins provides insight into Tobacco rattle virus replication complexes and reveals Arabidopsis DRB2 to be a wide-spectrum antiviral effector. Plant Cell 33(11): 3402–3420

Chapter 14

Analysis of Plant Virus-Induced Immunity by Using Viral-Derived Double-Stranded RNA in *Arabidopsis thaliana*

Penghuan Rui, Jianping Chen, Fei Yan, and Guanwei Wu

Abstract

Pattern-triggered immunity is the first line of defense against infection by pathogens such as bacteria and fungi in plants, and this mechanism remains poorly defined in plant viruses. Double-stranded RNA (dsRNA) is an intermediate in the replication of plant RNA viruses, and is considered to be a conserved structure of plant viruses similar to pathogen-associated molecular pattern. Whether dsRNA is the elicitor that activates plant immunity in response to virus infection remains obscure. In this method, we use the cDNA of turnip mosaic virus genome as the template to *in vitro* synthesis of viral dsRNA and examine whether viral dsRNA could activate plant immunity in *Arabidopsis thaliana*, including MAPK kinase cascade and reactive oxygen burst. In order to provide some references for researchers studying dsRNA in terms of research methodology and experimental methods, we use western blot to measure MAPK kinase cascade and luminol-based assay to measure ROS burst in *Arabidopsis thaliana* treated by viral dsRNA.

Key words Double-stranded RNA, Turnip mosaic virus, Reactive oxygen species, Plant immunity, MAPK kinase cascade, *Arabidopsis thaliana*

1 Introduction

Pathogen-associated molecular patterns (PAMPs) are the conserved patterns of pathogens [1, 2]. Plant hosts are able to broadly recognize all pathogens presenting this molecular pattern and mount an immune response that leads to broad-spectrum host resistance to pathogens. The immunity conferred by PAMPs, called pattern-triggered immunity (PTI), is therefore considered to be the first layer of active defense against pathogens in plants [3]. To date, all PAMP-sensing receptor proteins identified in plants are membrane-bound receptors. Upon recognition of PAMPs by these receptors, active receptor kinase complexes are formed at the plasma membrane, thereby activating phosphorylation events

Xiaofei Cheng and Guanwei Wu (eds.), *Double-Stranded RNA: Methods and Protocols*, Methods in Molecular Biology, vol. 2771, https://doi.org/10.1007/978-1-0716-3702-9_14,

that trigger downstream signaling, such as ion flux across the plasma membrane, reactive oxygen species production, activation of the mitogen-activated protein kinase (MAPK) cascade, hormonal signaling, and ultimately defense gene expression and defense metabolites and protein production [4, 5]. PAMPs are widely identified on microorganisms such as bacteria and fungi, and more recently on insects and nematodes [6, 7]. For plant viruses, until recently, PTI response is still largely unknown and no viral PAMPs have been identified.

It shows that the impaired resistance of *bak1*, *serk4*, and SERK-associated *nik1* plants to virus hints at a role for PTI-associated kinases in antiviral defense, and suggests that virus or virus-induced elicitations are present in virus-infected cells [8, 9]. However, these elicitors have not been identified to date either. Plant viruses are strictly obligate intracellular parasitic pathogens that typically enter the host cytoplasm directly through their insect vectors. In contrast to the case of microorganisms, it seems unlikely that structural features of virions can be sensed because of the large structural diversity of viruses. Therefore, it is difficult to imagine that they exist in a conserved structure such as PAMPs-like and are able to be recognized by receptors with recognition domains in the apoplast. In contrast, for viruses, conserved features that may represent viral PAMPs appear when viruses replicate within the host cell at the early infection stage [4].

DsRNA is a common pattern formed during replication of RNA and DNA viruses [10–12]. RNA-based pathogen-associated molecular patterns (PAMPs) are well-known inducers of immunity in animals [13, 14], and dsRNA can also induce immunity to viral and bacterial infections in plants, eliciting plant antiviral defense responses and activating typical PTI responses such as mitogen-activated protein kinase activity [15–17]. Thus, dsRNA may be recognized as bona fide PAMPs and induce plant immunity in plants.

Turnip mosaic virus (TuMV) has the widest host range among the potyviruses, infecting at least 318 species of 156 genera from 43 plant families, and one of its optimal hosts is *Arabidopsis thaliana* [18]. Therefore, TuMV is of great scientific research value and has attracted the attention of many plant virologists. Here, we use *in vitro* synthesized dsRNA using the complementary DNA of TuMV genome [19] as the template to treat *Arabidopsis thaliana*, aiming to verify whether dsRNA can induce the typical PTI response in plants, including the activation of MAPK kinase cascade and the generation of ROS, and to provide valuable research methods and experimental methods for reference.

2 Materials

2.1 In Vitro Synthesis of dsRNA

1. PCB301-TuMV-GFP plasmid [20] (*see* **Note 1**).
2. Generation of templates: 2×PrimeSTAR GXL Premix (Takara), 10 μM forward primer (T7 promoter): 5'-CGAAATTAATAC GACTCACTATAGGG caaacccagcagaatcggtg-3', 10 μM reverse prime (T7 promoter): 5'-CGAAATTAATACGACT CACTATAGGGgtcgccgtccatcatcaccc-3'.
3. NanoDrop One (Thermo Fisher Scientific).
4. Electrophoresis apparatus (BIO-RAD).
5. Mastercycler nexus GSX1 (Eppendorf).
6. HiPure gel pure DNA mini kit (Magen).
7. In vitro transcription: PCR product, NTP buffer mix (20 mM each NTP) (NEB), T7 RNA polymerase mix (NEB), RNase-free water, LiCl Solution (7.5 M LiCl, 10 mM EDTA) (NEB).
8. Agarose gel imaging system GelRed (Tanon).

2.2 Preparation of **Arabidopsis thaliana** *Leaf Discs*

1. Plants: *Arabidopsis thaliana* ecotype Col-0, 5- to 6-week-old.
2. Greenhouse: *Arabidopsis thaliana* plants are kept under 14 h/10 h light/dark periods at an average temperature of 22 °C in a growth chamber equipped with Philips LED lights.
3. Hole punch (1.0 cm in diameter).
4. Laboratory vacuum pump (Titan).
5. 96-well luminometer plate.
6. Aluminum foil.
7. Petri dishes.
8. Sterile absorbent paper.
9. Sterile filter paper.

2.3 Protein Extraction and SDS–Polyacrylamide Gel Electrophoresis

1. Extraction buffer: 0.01M PBS, 2% Na_2SO_3, 0.05% NP-40, 0.1% β-mercaptoethanol (*see* **Note 2**).
2. Loading buffer: 300 mM Tris–HCl (pH 6.8), 0.1% bromophenol blue, 8% (w/v) SDS, 25 % β-mercaptoethanol, 40% (v/v) glycerol.
3. Resolving gel: 1.5 M Tris–HCl, pH 8.8, 10% (w/v) SDS, 30% acrylamide/Bis solution (29:1), 10% Ammonium persulfate solution (APS), N, N, N, N′-Tetramethyl-ethylenediamine (TEMED).
4. Stacking gel: 0.5 M Tris–HCl, pH 6.8, 10% (w/v) SDS, 30% acrylamide/bis solution (29:1), 10% APS, TEMED.
5. Running buffer: Dissolve 1 g SDS, 14.4 g glycine and 3.03 g tris base in 900 mL water and then set the volume to 1 L.
6. Prestained Protein Marker, 14–120 kDa (TransGen Biotech).

2.4 Western Blot

1. Transfer buffer: isopropanol, PVDF membrane transfer buffer (GeneScript), ddH_2O.
2. Tris-buffered saline with Tween (TBS-T) (1×): 0.25 M NaCl, 3 mM Tris–HCl, pH 7.5, 1% (v/v) Tween 20.
3. Polyvinylidene fluoride membrane (PVDF, Sigma).
4. Methyl alcohol.
5. Blocking buffer: 5% (w/v) non-fat milk powder in 1× TBS-T.
6. Primary antibody: P-p44/42 MAPK (D13.14.4E) XP (R) Rabbit mAb (Cell Signaling Technology).
7. Secondary antibody: anti-rabbit IgG conjugated to HRP (TransGen Biotech).
8. Color development reagent: HRP conjugate substrate kit (Sigma).
9. Trans-Blot eBlot™ L1 (GeneScript).
10. Amersham imager 680 (GE Healthcare).
11. Coomassie brilliant blue staining solution (100 mL): 0.25g R250, 45 mL methyl alcohol, 10 mL glacial acetic acid, 45 mL ddH_2O.
12. Coomassie brilliant blue decolorization solution (100 mL): 45 mL methyl alcohol, 45 mL ddH_2O,10 mL glacial acetic acid.

2.5 Luminol-based Assay

1. Luminol (Sigma).
2. Horseradish peroxidase (Sigma).
3. Sterile distilled water.
4. Dimethyl sulfoxide (DMSO) (Sigma).
5. 1 mM flg22 (QRLSTGSRINSAKDDAAGLQIA) (Sigma).
6. GloMax®-Multi+ Detection System with dual injectors (Promega).
7. Poly(I:C) (Sigma).
8. 1×PBS: 8 g NaCl, 200 mg KCl, 1.44 g Na_2HPO_4, 240 mg KH_2PO_4 were dissolved in 900 mL of water, adjusted to PH 7.4, and then the volume is fixed to 1 L.

3 Methods

3.1 In Vitro Synthesis of dsRNA

1. Generate PCR products using respective reverse primer in combination with the universal dsRNA forward primer for generating a template for dsRNA (T7 promoter on each side) generation. PCR products are generated by amplifying fragments from PCB301-TuMV-GFP plasmid with the following

reagents: 25 μL PrimeSTAR GXL Premix (2×), 2.5 μL 10 μM forward primer and 10 μM reverse primer, respectively, 1 μL template, and 19 μL ddH_2O. In vitro amplification is performed according to the following procedure: 98 °C, 2 min; 98 °C, 15 s; 54 °C, 20 s; 68 °C, 1 min/1 kb (cycle for 35 times); 68 °C, 8 min.

2. Purification of PCR products: The PCR products need to be separated by agarose gel electrophoresis, and then the target band is cut off and purified with a gel purification kit. Purification of the product requires measurement of its concentration.
3. PCR products obtained from the above reactions are directly used for in vitro transcription. Following step: total volume of 20 μL containing 1 μg PCR product, 10 μL NTP buffer, 2 μL T7 RNA polymerase mix, RNase-free water. The reaction mix is incubated for 3 h at 37 °C.
4. Treat the dsRNA with DNase and RNase to remove residual DNA and ssRNA, and analyze the RNA samples by agarose gel electrophoresis (1.0% agarose in TAE buffer) using GelRed for UV-visualization.
5. After in vitro transcription, 30 μL RNase-free water and 25 μL LiCl solution are added to the reaction mix, followed by incubation at –20 °C for 30 min.
6. The precipitate is centrifuged at 4 °C for 15 min, washed with 75% ethanol, and resuspended in RNase-free water to obtain dsRNA. Determine the concentration of dsRNA spectrophotometrically at 260 nm.

3.2 Preparation of *Arabidopsis thaliana* Leaf Discs

1. Appropriate leaves are selected and harvested from at least two different plants and cut off from the petiole with a scalpel (*see* **Note 3**). The harvested leaves are gently washed with sterile distilled water (Fig. 1a) (*see* **Note 4**).

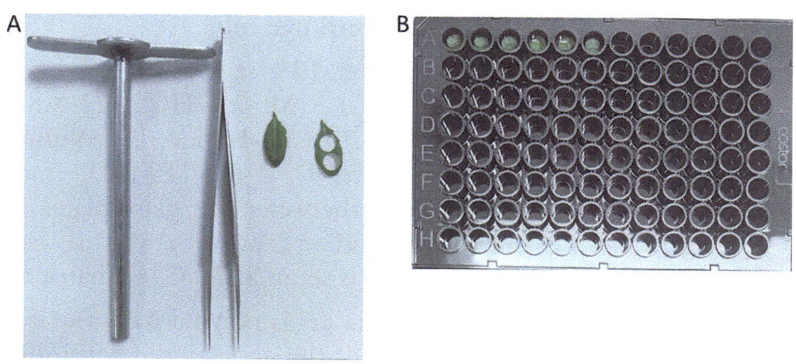

Fig. 1 Illustration of the preparation of leaf discs. (**a**) A paradigm for cutting Arabidopsis leaves to obtain leaf discs. From left to right, a hole punch, a needle-tipped tweezer, an Arabidopsis leaf, and a cut Arabidopsis leaf are displayed. (**b**) The picture shows a schematic of the leaf disc placed into a 96-well plate

2. The sterile water on the leaf surface is blotted dry with sterile absorbent paper, and then the leaves are placed on a new sterile filter paper. The leaf discs are sampled with a hole punch (1.0 cm in diameter) (*see* **Note 5**).
3. The obtained leaf discs are placed in Petri dishes containing sterile water and covered with aluminum foil, rinsed in the dark condition for 2 h, and the sterile water is changed every 30 min (*see* **Note 6**).
4. Using flat tweezers, one leaf per well is gently placed into a 96-well photometer plate containing 200 μL of sterile distilled water per well (Fig. 1b). Plates are covered with aluminum foil and incubated overnight in a growth chamber at 22 °C.

3.3 Poly(I:C) and dsRNA Treatment

1. Leaf discs incubated overnight are treated with RNase-free water, dsRNA, PBS, and poly(I:C) (*see* **Note 7**), respectively, and 1 μM poly(I:C) and dsRNA are added to each well, as well as equal volumes of RNase-free water and PBS as control treatments. Six treated leaf discs are treated as one biological replicate.
2. After the addition of the elicitors or the controls, the leaf discs are vacuum-infiltrated for 10 min. Samples are collected after an additional 20 min of incubation (*see* **Note 8**).
3. Samples from different treatments are collected separately, blotted dry with sterile absorbent paper, and frozen in liquid nitrogen for subsequent experiments.

3.4 Protein Extraction and SDS–Polyacrylamide Gel Electrophoresis

1. Samples frozen in liquid nitrogen are ground to powder, 100 μL of protein extraction buffer and 25 μL of loading buffer are added and mixed by shaking, followed by a water bath at 95 °C for 10 min.
2. The protein extract, which has completed denaturation, is cooled on ice, centrifuged at 12000 rpm for 5 min at 4 °C, and the supernatant is transferred to a 2 mL microcentrifuge tube.
3. Prepare a 12% separating gel by mixing 1.9 mL ddH_2O, 1.3 mL 1.5 M Tris–HCl, pH 8.8, 50 μL 10% (w/v) SDS, 1.7 mL 30% acrylamide/Bis solution (29:1), 50 μL 10% APS (*see* **Note 9**), 2 μL TEMED. Immediately pour 5 mL gel mixture between the glass plates seated on the gel casting stand and overlay with 500 μL isopropanol (*see* **Note 10**). The gel is set in a 37 °C incubator for about 15 min.
4. Once the gel is polymerized, the isopropanol is decanted and the top of the gel is rinsed with ddH_2O. Prepare the stacking gel by mixing 3.4 mL ddH_2O, 0.63 mL 0.5 M Tris–HCl, pH 6.8, 50 μL 10% (w/v) SDS, 0.83 mL 30% acrylamide/Bis

solution (29:1), 50 μL 10% APS, 5 μL TEMED. Immediately pour the stack on top of the gels and insert the 10-well combs. The gel will set in a 37 °C incubator for about 20 min (*see* **Note 11**).

5. After the gel plate is properly mounted, 1 L of running buffer is added, the comb is pulled out, and 10 μL sample is loaded.
6. Connect the cable to the power supply and maintain a constant operating voltage of 80 V until the dye front reaches the bottom of the gel. This process takes approximately 2 h.

3.5 Western Blot

1. Soak the size of 8.5 × 4.5 cm PVDF membrane in methyl alcohol for 1 min (*see* **Note 12**).
2. Prepare the transfer buffer by mixing 0.5 L isopropanol, 1 L PVDF membrane transfer buffer, 3.5 L ddH_2O.
3. At the end of the electrophoresis, the gel is carefully removed. Then install the blotting sandwich correctly according to the manufacturer's instructions.
4. Place the cassette in module of Trans-Blot (*see* **Note 13**), click start to run the program.
5. At the end of the procedure run, the membranes are carefully removed and incubated in blocking buffer for 30 min.
6. Dilute anti-MAPK rabbit antibody (usually 1:5000) in 20 mL blocking buffer and incubate the membrane overnight at 4 °C (*see* **Note 14**).
7. Membrane is rinsed four times with TBS-T for 5 min each (*see* **Note 15**).
8. Dilute anti-Rabbit IgG conjugated to HRP antibody (usually 1:10,000) in 20 mL blocking buffer and incubate the membrane for 1 h at room temperature.
9. Membrane is rinsed eight times with TBS-T for 5 min each.
10. 100 μL A solution and 100 μL B solution of HRP conjugate substrate are mixed to be evenly applied to the membrane. Scanning imaging is performed by Amersham Imager 680. Please check the instructions for how to use the system.
11. After scanning image, the membrane is stained with Coomassie brilliant blue for 5 min (*see* **Note 16**), and then decolorized with decolorization solution for 10 min.
12. The results of our assay indicate that TuMV-derived dsRNA can significantly activate the MAPK kinase cascade compared with RNase-free water treatment (Fig. 2). Poly(I:C) as a positive control activates the MAPK kinase cascade at a higher level than the control (PBS), which is consistent with previous findings [16].

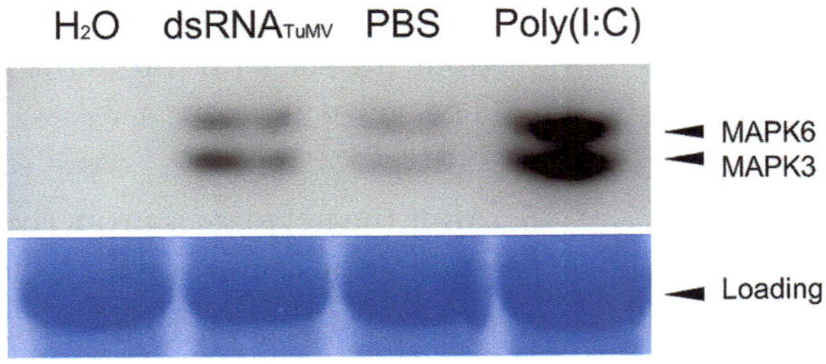

Fig. 2 TuMV-derived dsRNA induced the phosphorylation of MAPK3/6. The dsRNA and different control treatments were performed on Col-0 leaf discs, followed by western blot using a primary antibody of P-p44/42 MAPK (D13.14.4E) XP (R) Rabbit mAb

3.6 Luminol-Based Assay

1. Luminol stock solution is prepared by dissolving 15 mg of luminol in 1 mL of dimethyl sulfoxide and vortexing until the solution turns to a uniform light green color. The horseradish peroxidase stock solution is prepared by dissolving 10 mg of horseradish peroxidase in 1 mL of sterile distilled water and vortexing until the solution turns to a uniform brown color. 1 mM flg22 stock solution is prepared by dissolving 2.27 mg of flg22 in 1 mL of sterile distilled water, by vortexing until the solution becomes transparent (*see* **Note 17**).
2. Prepare a luminol/peroxidase working solution contaning 0.5 μL luminol stock solution and 0.5 μL horseradish peroxidase stock solution up to 200 μL with sterile distilled water. Gently and rapidly replace the sterile distilled water with an equal volume of solution containing 1 μM of flg22 or dsRNA in each well containing leaf discs, which are then vacuum-infiltrated as described in Subheading 3.3, **step 2**. Use leaf discs by replacing the water with the same sterile distilled water or PBS for flg22 or dsRNA suspension as a negative control (*see* **Note 18**).
3. Immediately manually place the plate in the GloMax®-Multi+ Detection System, inject 50 μl of the luminol/peroxidase working solution per well, and click start to run the program according to the manufacturer's instructions. Perform the luminescence detection assay for 40 min with a signal interation time of 2 min.
4. The results of the analysis indicate that virus-derived dsRNA is unable to induce ROS burst throughout the testing process, whereas flg22, a positive control, is able to induce strong ROS burst in *Arabidopsis thaliana* leaves within 10 min (Fig. 3).

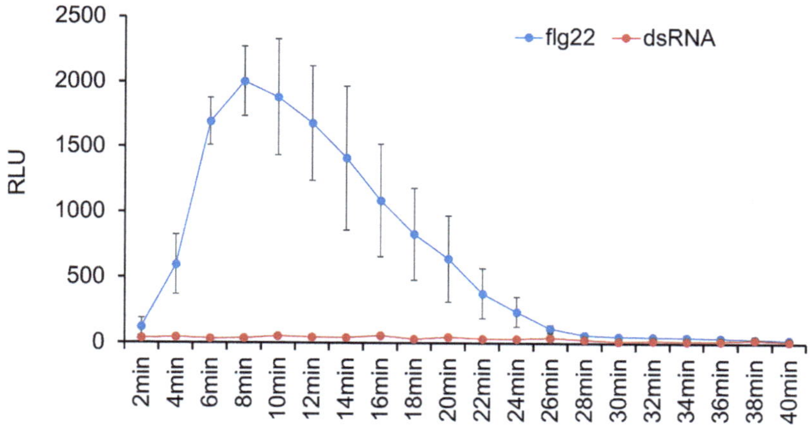

Fig. 3 TuMV-derived dsRNA does not induce ROS Burst. Col-0 leaf discs were treated with dsRNA or flg22, respectively, followed by a luminol-based assay. RLU, relative luminescence units. Mean values and error bars (standard deviation) were obtained for each time point for 6 replicates (leaf discs) per treatment

4 Notes

1. PCB301-TuMV-GFP (Accession number: NC_002509) plasmid.
2. All reagents or buffers are prepared using sterile ultrapure water at room temperature unless otherwise stated. Protective measures should be taken when configuring all reagents. The extraction buffer needs to be placed at 4 °C for preservation.
3. In order to make the best experimental condition of *Arabidopsis thaliana*, it needs to be placed in a constant temperature incubator at 22 °C, watered and dewormed regularly, and not allowed to suffer from any biotic and abiotic stresses. To make leaf discs, the third to fifth leaf from *Arabidopsis thaliana* is selected. When picking leaves, avoid other mechanical damage to the leaves.
4. The harvested *Arabidopsis thaliana* leaves need to be lightly washed with running sterile water to remove dust and possible pathogenic bacteria from the leaf surface and try to ensure that the leaves are free from any irritation.
5. Try to keep the cork borer sharp enough to minimize damage to the leaves. The whole process is carried out as far as possible in a sterile or relatively clean environment to avoid stimulation of leaves by pathogens in the environment. The cut leaf discs should be placed in sterile water immediately to avoid prolonged exposure to air, which can cause leaf disc dehydration and thus suffer stress.

6. The leaves do not need to be completely submerged in water during this procedure. However, it is important that they are in good contact with aqueous media to prevent dehydration and to remove secondary metabolites produced upon wounding.
7. To create a more suitable environment for the leaf discs, poly(I: C) is dissolved with sterilized PBS buffer, and sterilized PBS buffer is treated as a control. Poly(I:C) should be stored in aliquots after use and can be stored at −20 °C for 6 months.
8. It is recommended to add the elicitors or the controls with a shotgun to ensure that all treatments are added in a very short time and to reduce experimental errors. In order to allow the elicitors or the controls to fully enter the leaf disc, vacuum suction filtration can be used to achieve better treatment results. Ensure that the leaf discs are completely immersed in the solution.
9. 10% APS solution is unstable and can only be stored for 1 week even at 4 °C. It is recommended to configure it before use, otherwise it may cause the separating gel and stacking gel to fail to solidify. The preparation of resolving gel and stacking gel needs to be done in a fume hood.
10. Before pouring the gel mixture, it is necessary to check the airtightness of the whole device, otherwise it will lead to gel leakage. The covering of isopropanol above the separating gel is to allow the gel to be flatter and to avoid the destruction of the integrity of the gel by bubbles produced by SDS. Here we use isopropyl alcohol instead of water for better results.
11. Compared with room temperature, placing the gel at 37 °C accelerates the gel solidification without affecting the use of the gel.
12. The PVDF membrane needs to be activated with methanol before transferring, and then rinsed with the transfer buffer.
13. The system is a rapid membrane transfer system from GeneScript, which can complete protein transfer in 15 min. The rapid membrane transfer system has a special membrane transfer buffer, which is different from the wet membrane transfer buffer, do not mix, otherwise it will lead to experiment failure, more serious will lead to instrument errors. Please check the instructions for how to use the system.
14. Primary antibody incubation is recommended to be placed at 4 °C overnight for static incubation. It can effectively reduce the nonspecific binding of antibody and avoid strong background noise during scanning imaging.
15. Since TBS-T contains Tween 20, it needs to be stored away from light. It is recommended to use it as soon as possible after

configuration, otherwise it will affect the effect of TBS-T cleaning PVDF membrane, resulting in higher noise of PVDF membrane during scanning imaging and affecting the experimental results.

16. The staining time of Coomassie brilliant blue is about 5–10 min, and staining with too long time will cause a relatively strong background. The whole staining process needs to be carried out in a fume hood because the staining solution contains many volatile chemical reagents that are harmful to health.
17. Luminol needs to be dissolved in DMSO, and it is easy to decompose in the light, so it needs to be stored in the dark. The unused Luminol needs to be stored in aliquots to −20 °C, and the validity period is 3 months. Horseradish peroxidase and flg22 can be dissolved in sterile water, which also needs to be stored in the dark, and can be stored in aliquots at −20 °C for 6 months.
18. It is necessary to be careful that ROS can be induced in a very short time after the addition of the elicitors, so the elicitors and controls should be added quickly and tested immediately. It is recommended that the test instrument is adjusted in advance to ensure that it can be operated immediately after the treatment is completed.

Acknowledgments

This research is funded by grants from the National Natural Science Foundation of China (3207016) to GW, the China Agriculture Research System of MOF and MARA (CARS-24-C-04) to FY, and the K. C. Wong Magna Fund in Ningbo University.

References

1. Couto D, Zipfel C (2016) Regulation of pattern recognition receptor signalling in plants. Nat Rev Immunol 16:537–552
2. Boller T, Felix G (2009) A renaissance of elicitors: perception of microbe-associated molecular patterns and danger signals by pattern-recognition receptors. Annu Rev Plant Biol 60:379–406
3. Jones JDG, Dangl JL (2006) The plant immune system. Nature 444:323–329
4. Amari K, Niehl A (2020) Nucleic acid-mediated PAMP-triggered immunity in plants. Curr Opin Virol 42:32–39
5. DeFalco TA, Zipfel C (2021) Molecular mechanisms of early plant pattern-triggered immune signaling. Mol Cell 81:3449–3467
6. Abdul Malik AN, Kumar SI, Nadarajah K (2020) Elicitor and receptor molecules: orchestrators of plant defense and immunity. Int J Mol Sci 21:963
7. Manosalva P, Manohar M, Von Reuss SH et al (2011) Conserved nematode signalling molecules elicit plant defenses and pathogen resistance. Nat Commun 6:7795
8. Kørner CJ, Klauser D, Niehl A et al (2013) The immunity regulator BAK1 contributes to resistance against diverse RNA viruses. Mol Plant Microbe In 26:1271–1280

9. Zorzatto C, Machado JPB, Lopes KVG et al (2015) NIK1-mediated translation suppression functions as a plant antiviral immunity mechanism. Nature 520:679–682
10. Paude DB, Sanfaçon H (2018) Exploring the diversity of mechanisms associated with plant tolerance to virus infection. Fron Plant Sci 9: 1575
11. Pumplin N, Voinnet O (2013) RNA silencing suppression by plant pathogens: defence, counter-defence and counter-counter-defence. Nat Rev Microbiol 11:745–760
12. Ramesh SV, Sahu PP, Prasad M et al (2017) Geminiviruses and plant hosts: A closer examination of the molecular arms race. Viruses 9: 256
13. Chow KT, Gale MJ, Loo YM (2018) RIG-I and other RNA sensors in antiviral immunity. Annu Rev Immunol 36:667–694
14. Vabret N, Bhardwaj N, Greenbaum BD (2017) Sequence-specific sensing of nucleic acids. Trends Immunol 38:53–65
15. Lee B, Park YS, Lee S et al (2016) Bacterial RNAs activate innate immunity in *Arabidopsis*. New Phytol 209:785–797
16. Niehl A, Wyrsch I, Boller T et al (2016) Double-stranded RNAs induce a pattern-triggered immune signaling pathway in plants. New Phytol 211:1008–1019
17. Huang CP, Sede A, Elvira-González L et al (2023) DsRNA-induced immunity targets plasmodesmata and is suppressed by viral movement proteins. Plant Cell. https://doi.org/10.1093/plcell/koad176
18. Nellist CF, Ohshima K, Ponz F et al (2022) Turnip mosaic virus, a virus for all seasons. Ann Appl Biol 180:312–327
19. Voloudakis AE, Holeva MC, Sarin LP et al (2015) Efficient double-stranded RNA production methods for utilization in plant virus control. Methods Mol Biol 1236:255–274
20. Wu GW, Jia ZX, Ding KD et al (2022) Turnip mosaic virus co-opts the vacuolar sorting receptor VSR4 to promote viral genome replication in plants by targeting viral replication vesicles to the endosome. PLoS Pathog 18: e1010257

Chapter 15

Identification of Mycoviruses by dsRNA Extraction

Yanfei Wang, Clement Nzabanita, and Lihua Guo

Abstract

Mycoviruses exist in all major groups of fungi. With the continuous development of science and technology, the methods of studying viruses are constantly updated, and progressively mycoviruses have been discovered where most of these viruses are RNA viruses. Therefore, double-stranded RNA has traditionally been used as the hallmark of RNA mycovirus detection. This report describes in detail the method of mycovirus identification using extraction of dsRNA. Besides, extraction of viral dsRNA, and the assembly methods of viral genome and identification of virus type are presented.

Key words Mycoviruses, dsRNA viruses, dsRNA extraction, Viral genome assembly, Viral classification

1 Introduction

Fungal viruses (mycoviruses) exist in all major groups of fungi [1]. The first mycovirus was discovered when investigating the causing agent of the mushroom die-back disease [2], after which the knowledge concerning mycoviruses has gradually expanded. With the continuous development of next-generation sequencing (NGS) technologies, the number of known mycoviruses has increased rapidly due to extensive researches on plant pathogenic fungi [1]. According to the latest ICTV taxonomy report, the currently identified 206 mycoviruses are classified into 23 families [3–7], including nine families with positive single-stranded RNA (+) ssRNA genomes (*Barnaviridae*, *Endornaviridae*, *Alphaflexiviridae*, *Deltaflexiviridae*, *Gammaflexiviridae*, *Narnaviridae*, *Mitoviridae*, *Botourmiaviridae,* and *Hypoviridae*), two families with negative single-stranded RNA (−) ssRNA genomes (*Mymonaviridae* and *Phenuiviridae*), nine families with double-stranded RNA (dsRNA) genomes (*Chrysoviridae*, *Curvulaviridae*, *Megabirnavir-*

Xiaofei Cheng and Guanwei Wu (eds.), *Double-Stranded RNA: Methods and Protocols*, Methods in Molecular Biology, vol. 2771, https://doi.org/10.1007/978-1-0716-3702-9_15,

idae, *Partitiviridae*, *Quadriviridae*, *Reoviridae*, *Totiviridae*, *Polymycoviridae*, and *Botybirnaviridae*), one family with single-stranded DNA (ssDNA) genomes (*Genomoviridae*) [8, 9], and two families with reverse transcribing single-stranded RNA (RT-ssRNA) genomes (*Metaviridae* and *Pseudoviridae*).

With increasingly serious environmental pollution and food safety problems caused by chemical fungicides, it is urgent to develop green, environmentally friendly and efficient disease prevention and control technology [10]. The use of mycoviruses with biocontrol potential to control fungal diseases is not only an efficient method but also environmentally friendly. Mycoviruses can induce hypovirulence of host pathogenic fungi and have gained considerable attention as potential biological control tools [11–13]. The most notable example of the successful deployment of hypovirulence-induced viruses to control plant pathogens in the field is the application of Cryphonectria hypovirus 1 (CHV-1) to chestnut disease control in Europe [14, 15]. Subsequently, Zhang et al. (2020) used DNA mycovirus Sclerotinia sclerotiorum hypovirulence-associated DNA virus 1 (SsHADV-1) to transform its host (*Sclerotinia sclerotiorum*) from a typical necrotic pathogen to a beneficial endophytic fungus, thereby promoting plant growth and enhancing disease resistance [8, 16, 17]. At present, despite numerous kinds of mycoviruses, few are used as biocontrol agents [10]. Therefore, it is necessary to continue to excavate a large number of mycoviruses to provide a rich resource for fungal viruses as biological control agents.

Most of the mycoviruses discovered so far are RNA viruses. RNA viruses have ssRNA or dsRNA as their genomes [5]. In infected cells, ssRNA viruses accumulate a significant amount of dsRNAs as their replicative intermediate. Therefore, dsRNA has traditionally been used as the hallmark of RNA mycovirus detection [18]. The rapid and specific extraction method for dsRNA from fungal cells and the conventional detection method of agarose gel electrophoresis (AGE) have accelerated a large-scale of mycovirus screening [19–22]. Many procedures for isolation and detection of dsRNA have been established, such as chromatographic separation (e.g., on hydroxyapatite, diethylaminoethyl cellulose, or epichlorohydrin triethanolamine cellulose), enzymatic degradation of ssRNA and DNA, and salt fractionation using lithium chloride [23]. Chromatography using CF-11 cellulose powder may be the most practical method, and has been widely utilized for the purification of viral dsRNA [21, 23]. This report introduces in detail the method of mycovirus identification using dsRNA extraction. In addition, extraction of viral dsRNA, and assembly methods of viral genome and identification of virus type are also elucidated.

2 Materials

2.1 Chemical and Reagents

1. D-Glucose (e.g., Thermo Fisher Scientific, Catalog No. D16-500).
2. Agar (e.g., BD Difco, Catalog No. DF0812-07-1).
3. Sodium chloride (NaCl; e.g., Fisher Chemical, Catalog No. S271-1).
4. Tryptone (e.g., Fisher BioReagents, Catalog No. BP1421).
5. Yeast extract powder (e.g., MP Biomedicals, Catalog No. ICN10330391).
6. Tris base (e.g., Fisher BioReagents, Catalog No. BP152-10).
7. Ethylenediaminetetraacetic acid (EDTA; e.g., Thermo Fisher Scientific, Catalog No. S311-3).
8. Ethanol absolute (e.g., Fisher BioReagents, Catalog No. BP2818100).
9. Sodium acetate anhydrous (e.g., Fisher BioReagents, Catalog No. BP333-1).
10. Sodium Dodecyl Sulfate (SDS; e.g., Fisher BioReagents, Catalog No. BP166-500).
11. Cellulose powder CF-11 (e.g., Merck Sigma Aldrich, Catalog No. 9004-34-6).
12. Phenol/Tris-saturated (e.g., Thermo Fisher Scientific, Catalog No. AC327125000).
13. M-MLV Reverse Transcriptase (e.g., Promega, Catalog No. M1701).
14. TransScript One-Step RT-PCR SuperMix (e.g., Transgene, Catalog No. AT411-02).
15. Terminal deoxynucleotidyl transferase (TdT), Recombinant (e.g., Promega, Catalog No. M1871).
16. T4 RNA ligase (e.g., Promega, Catalog No. M1051).
17. Superscript II reverse transcriptase (e.g., Invitrogen, Catalog No. 18064014).
18. pMD18-T vector (e.g., Takara, Catalog No. 6011).
19. *Escherichia coli* strain DH5α.

2.2 Culture Medium

1. PDA (potato dextrose agar) solid medium: Put 200 g peeled potatoes and 1 L ddH$_2$O into a pot, boil for 15 min, filter with three layers of gauze, add 20 g glucose, fill ddH$_2$O to 1 L, divide 300 mL into a 500 mL conical flask, and add 4.5 g agar to each flask, steam sterilize at 121 °C for 20 min.

2. LB liquid medium: Weigh 10 g Tryptone, 5 g Yeast extract, and 10 g NaCl, dissolve them in 1 L deionized water, and steam sterilize at 121 °C for 20 min.
3. LB solid medium: Weigh 10 g Tryptophan, 5 g Yeast extract, and 10 g NaCl, dissolve them in 1 L deionized water and divide 300 mL into a 500 mL conical flask, add 4.5 g agar in each flask and steam sterilize at 121 °C for 20 min.

2.3 Buffers

1. STE buffer: 0.1 mol/L NaCl, 10 mmol/L Tris, 1 mmol/L EDTA, pH = 8.0.
2. STE buffer containing 16% ethanol: Measure 16 mL of anhydrous ethanol in a measuring cylinder, add 84 mL of STE buffer, fix the volume to 100 mL.
3. 10% SDS: 10 g SDS dissolved in 100 mL of double-distilled water.
4. 3.0 M sodium acetate buffer: 24.61 g NaAc dissolved in 90 mL of double-distilled water, adjust the pH to 5.2, add double-distilled water to the volume of 100 mL.

3 Methods

3.1 Extraction of Viral dsRNA

Extract viral dsRNA using the method of CF-11 cellulose chromatography [24]. The specific steps are as follows:

1. Place mycelial plugs of activated strain onto a PDA plate overlaid with cellophane membranes and culture at 25 °C for 4 days in the dark.
2. Collect the mycelium in a 50 mL centrifuge tube, frozen and ground to a powder in liquid nitrogen using a sterile pestle and mortar.
3. Resuspend the powder in twice the volume (w/v) of 1xSTE buffer, equal volume(v/v) of Tris-saturated phenol and 2% of 10% SDS.
4. Shake the suspension at room temperature for 1 h.
5. Centrifuge for 20 min at 10,000 rpm, collect the upper phase, and add isovolumetric chloroform.
6. Shake at room temperature for 10 min, centrifuge for 10 min at 12,000 rpm.
7. Collect the upper phase, add 4% CF-11 cellulose, and add anhydrous ethanol until the final concentration of 16%.
8. Gently mix at room temperature for 30 min, centrifuge for 3 min at 12,000 rpm, and collect the pellet.

9. Resuspend the pellet in 20 mL of STE buffer containing 16% v/v ethanol, gently mix at room temperature for 1 min, centrifuge for 3 min at 12,000 rpm, and collect the pellet.
10. Repeat the previous step one more time.
11. Resuspend the pellet in 20 mL of STE buffer, gently mix at room temperature for 1 min.
12. Add 1/10 volume of 3.0 M sodium acetate and 2.5 times the volume of ethanol. Store for 1 h at −80 °C to precipitate the dsRNA.
13. Centrifuge at 12,000 rpm for 15 min at 4 °C. Pour off the ethanol and place the tubes upside down for about 15 min.
14. Add 200 μL of DEPC-treated ddH_2O to resuspend the dsRNA. Store at −80 °C.

3.2 Assembly of the Viral Genome

1. cDNAs clone using random primers.
   (a) Use the extracted dsRNA of the virus as a template; perform reverse transcription with random primers (26-mer, 5′-GACGTCCAGATCGCGAATTCNNNNNN-3′). The specific steps are as follows: mix 2.0 uL dsRNA, 1 uL random primers and 13.75 uL ddH_2O; denature at 95 ° C for 10 min and immediately chill on ice for 5 min. Then add 5.0 uL 5× reaction buffer, 1.25 uL dNTP, 1uL RNase inhibitor and 1 uL M-MLV Reverse Transcriptase (Promega, USA). Mix and react at 42 °C for 1 h and 75 °C for 10 min. Use the cDNA synthesized in the above step as the template for PCR amplification with the random primer (20-mer, 5′- GACGTCCAGATCGCGAATTC - 3′). Separate and purify the PCR product using a gel extraction kit (Sigma) on a 1% agarose gel. Ligate the products to the PMD18-T vector and transform it into *Escherichia coli* strain DH5α. Select the positive clones for sequencing.
2. cDNAs clone using specific primers.
   (a) Design the specific primers according to the obtained dsRNA sequences. Determine the sequences in the gaps between the cDNAs by RT-PCR using specific primers. Then perform RT-PCR as described in the previous step. Separate and purify the PCR product using a gel extraction kit (Sigma) on a 1% agarose gel. Ligate the products to the PMD18-T vector and transform it into *Escherichia coli* strain DH5α. Select the positive clones for sequencing.
3. Clone of termini of the dsRNA.
   (a) To obtain the sequences of the virus, clone the 5′ end and 3′ end sequences using the RACE (rapid amplification of cDNA ends) cloning strategy.

(b) 5′-RACE is a ployC tail attached to the 3′ end of the retro transcriptional product cDNA using terminal deoxynucleotidyl transferase (TdT), and then nested PCR amplification using splice primers and sequence-specific primers. Specific experimental methods are as follows: mix 2.0 uL dsRNA, 1 uL sequence-specific primers and 15.5 uL ddH_2O; denature at 95 °C for 10 min and immediately chill on ice for 5 min. Then add 2.5 uL 10× reaction buffer, 2.5 uL 25 mM $MgCl_2$, 1 uL dNTP and 1 uL Reverse Transcriptase, mix and react at 42 °C for 1 h and 75 °C for 10 min. Then add 2.9 uL 10 × RNase H buffer and 1 uL RNase H to the above reaction system, and react at 37 °C for 30 min to remove the RNA of the hybrid molecules. Prepare the following reaction mixture for N-long tail: 10 uL cDNA, 2.5 uL dCTP, 6.5 uL ddH_2O and 5 uL 5× TdT buffer. Denature at 95 °C for 2 min and immediately chill on ice for 5 min, add 1uL TdT and incubate the mixture at 37 °C for 10 min. Using cDNA products as template, perform nested PCR using the 5′ abridged anchor primer (5′- GGCCACGCGTC GACTAGTACGGGIIGGGIIGGGIIG-3′) and a nested specific primer. Ligate the products to the PMD18-T vector and transform it into *Escherichia coli* strain DH5α. Select the positive clones for sequencing.

(c) A 3′ RNA Ligase-Mediated RACE (RLM-RACE) is that the 3′ terminus of each strand of dsRNA was ligated at 4–8 °C for 18 h with the 5′-end phosphorylated oligonucleotide 5′-GCATTGCATCATGATCGATCGAATTCTT TAGTGAGGGTTAATTGCC-(NH_2)-3′ using T4 RNA ligase. Denature and use the oligonucleotide-ligated dsRNA for the reverse transcription reaction with Superscript II reverse transcriptase and use 3 pmol of a primer with sequence complementary to the oligonucleotide (5′-GGCAATTAACCCTCACTAAAG-3′) for the RNA ligation. Treat the reaction product with RNase H as described above and amplify the cDNA with another primer complementary to the RNA ligation oligonucleotide (5′-TCACTAAAGAATTCGATCGATC-3′) and the sequence-specific primer corresponding to the 3′-terminal sequences of the dsRNA. Ligate the products to the PMD18-T vector and transform it into *Escherichia coli* strain DH5α. Select the positive clones for sequencing.

4. Assembly of complete viral genome sequence.

(a) Assemble and analyze the resulting sequences using DNAMAN software. Splice all the sequences obtained with DNAMAN software to obtain a complete virus genome sequence.

3.3 Sequencing and Phylogenetic Analyses

1. Analyze the open reading frames (ORFs) of virus genome sequence by DNAMAN version 8 software.
2. Align sequences by the ClustalW program, then analyze sequence homologs and conserved domains.
3. Perform sequence similarity searches in the National Center for Biotechnology Information (NCBI) BLAST program (Blastn, Blastp).
4. Construct the phylogenetic tree with the neighbor-joining method with 1000 bootstraps in MEGA 6 program.
5. Analyze the potential of RNA secondary structure using the UNAFold Web Server (http://unafold.rna.albany.edu).

4 Notes

1. The vessels used in the dsRNA extraction process should be treated with 0.1% DEPC aqueous solution at 37 °C for 12 h in advance, and then autoclaved at 121 °C for 30 min [25].
2. The cDNA cloned by random primers should be sequenced by selecting monoclonal sequences with different fragment sizes to obtain as many sequences of different regions of the viral genome as possible [25].
3. Virus-specific primers should be designed in two different fragments, while paying attention to the orientation of the primers [25].
4. When the sequence is assembled, there are at least three sequencing results for each site to ensure the accuracy of the sequence [25].

References

1. Ghabrial SA, Suzuki N (2009) Viruses of plant pathogenic fungi. Annu Rev Phytopathol 47: 353–384. https://doi.org/10.1146/annurev-phyto-080508-081932
2. Hollings M (1962) Viruses associated with a die-back disease of cultivated mushroom. Nature 196(4858):962–965. https://doi.org/10.1038/196962a0
3. Kondo H, Botella L, Suzuki N (2022) Mycovirus diversity and evolution revealed/inferred from recent studies. Annu Rev Phytopathol 60: 307–336. https://doi.org/10.1146/annurev-phyto-021621-122122
4. Ghabrial SA, Caston JR, Jiang D, Nibert ML, Suzuki N (2015) 50-plus years of fungal viruses. Virology 479-480:356–368. https://doi.org/10.1016/j.virol.2015.02.034
5. Myers JM, James TY (2022) Mycoviruses. Curr Biol 32(4):R150–R155. https://doi.org/10.1016/j.cub.2022.01.049
6. Kotta-Loizou I (2021) Mycoviruses and their role in fungal pathogenesis. Curr Opin Microbiol 63:10–18. https://doi.org/10.1016/j.mib.2021.05.007
7. Casas LL, Azevedo JL, Almeida LN, Costa-Neto PQ, Bianco RA, Pereira JO (2021) Mycoviruses infecting Colletotrichum spp.: a comprehensive review. Braz J Biol 83: e248975. https://doi.org/10.1590/1519-6984.248975
8. Yu X, Li B, Fu Y, Jiang D, Ghabrial SA, Li G, Peng Y, Xie J, Cheng J, Huang J, Yi X (2010) A geminivirus-related DNA mycovirus that confers hypovirulence to a plant pathogenic

fungus. Proc Natl Acad Sci U S A 107(18): 8387–8392. https://doi.org/10.1073/pnas.0913535107

9. Li P, Wang S, Zhang L, Qiu D, Zhou X, Guo L (2020) A tripartite ssDNA mycovirus from a plant pathogenic fungus is infectious as cloned DNA and purified virions. Sci Adv 6(14): eaay9634. https://doi.org/10.1126/sciadv.aay9634
10. Tonka T, Walterová L, Čurn V (2022) Biological control of pathogenic fungi: can mycoviruses play an important role? J Cent Eur Agric 23(3):540–551. https://doi.org/10.5513/jcea01/23.3.3618
11. Liu H, Wang H, Liao XL, Gao B, Lu X, Sun D, Gong W, Zhong J, Zhu H, Pan X, Guo L, Deng XW, Zhou Q (2022) Mycoviral gene integration converts a plant pathogenic fungus into a biocontrol agent. Proc Natl Acad Sci U S A 119(50):e2214096119. https://doi.org/10.1073/pnas.2214096119
12. Son M, Yu J, Kim KH (2015) Five questions about mycoviruses. PLoS Pathog 11(11): e1005172. https://doi.org/10.1371/journal.ppat.1005172
13. Das T, Prasad A, Dey A (2023) Mycoviral gene-incorporating phytopathogenic fungi: a biocontrol agent. Trends Plant Sci 28(8): 864–866. https://doi.org/10.1016/j.tplants.2023.05.005
14. Heiniger U, Rigling D (1994) Biological control of chestnut blight in Europe. Annu Rev Phytopathol 32(1):581–599. https://doi.org/10.1146/annurev.py.32.090194.003053
15. Anagnostakis SL (1982) Biological control of chestnut blight. Science 215(4532):466–471. https://doi.org/10.1126/science.215.4532.466
16. Zhang H, Xie J, Fu Y, Cheng J, Qu Z, Zhao Z, Cheng S, Chen T, Li B, Wang Q, Liu X, Tian B, Collinge DB, Jiang D (2020) A 2-kb mycovirus converts a pathogenic fungus into a beneficial endophyte for Brassica protection and yield enhancement. Mol Plant 13(10): 1420–1433. https://doi.org/10.1016/j.molp.2020.08.016
17. Yu X, Li B, Fu Y, Xie J, Cheng J, Ghabrial SA, Li G, Yi X, Jiang D (2013) Extracellular transmission of a DNA mycovirus and its use as a natural fungicide. Proc Natl Acad Sci U S A 110(4):1452–1457. https://doi.org/10.1073/pnas.1213755110
18. Chiba Y, Oiki S, Yaguchi T, Urayama SI, Hagiwara D (2021) Discovery of divided RdRp sequences and a hitherto unknown genomic complexity in fungal viruses. Virus Evol 7(1): veaa101. https://doi.org/10.1093/ve/veaa101
19. Okada R, Kiyota E, Moriyama H, Fukuhara T, Natsuaki T (2015) A simple and rapid method to purify viral dsRNA from plant and fungal tissue. J Gen Plant Pathol 81(2):103–107. https://doi.org/10.1007/s10327-014-0575-6
20. Khankhum S, Escalante C, de Souto ER, Valverde RA (2017) Extraction and electrophoretic analysis of large dsRNAs from desiccated plant tissues infected with plant viruses and biotrophic fungi. Eur J Plant Pathol 147(2): 431–441. https://doi.org/10.1007/s10658-016-1014-7
21. Arjona-Lopez JM, Telengech P, Jamal A, Hisano S, Kondo H, Yelin MD, Arjona-Girona I, Kanematsu S, Lopez-Herrera CJ, Suzuki N (2018) Novel, diverse RNA viruses from Mediterranean isolates of the phytopathogenic fungus, Rosellinia necatrix: insights into evolutionary biology of fungal viruses. Environ Microbiol 20(4):1464–1483. https://doi.org/10.1111/1462-2920.14065
22. Morris TJ, Dodds JA (1979) Isolation and analysis of double-stranded RNA from virus-infected plant and fungal tissue. Phytopathology 8:854–858. https://doi.org/10.1094/Phyto-69-854
23. Atsumi G, Sekine K-T, Kobayashi K (2015) A new method to isolate total dsRNA. In: Uyeda I, Masuta C (eds) Plant virology protocols: new approaches to detect viruses and host responses. Springer, New York, pp 27–37. https://doi.org/10.1007/978-1-4939-1743-3_3
24. Valverde RA, Nameth ST, Jordan R (1990) Analysis of double-stranded RNA for plant virus diagnosis. Plant Dis 74:255–258
25. Wang S, Kondo H, Liu L, Guo L, Qiu D. (2013) A novel virus in the family Hypoviridae from the plant pathogenic fungus Fusarium graminearum. Virus Res 174(1–2):69–77. https://doi.org/10.1016/j.virusres.2013.03.002. Epub 2013 Mar 14. PMID: 23499998

Chapter 16

Production of Double-Stranded RNA in Planta by a Potato Mop-Top Virus (PMTV)-Based Vector for Inducing Gene Silencing

Ye Liu, Yanlin Gao, Xiaofei Cheng, and Yanju Bai

Abstract

RNA silencing (also known as gene silencing) is an evolutionary conserved mechanism that is involved in regulating gene expression, suppressing mobile elements, and defensing virus infection. RNA silencing is triggered by double-stranded RNA via Dicer or Dicer-like riboendonucleases. DsRNAs are also the replication intermediates of all RNA viruses; as a result, plant RNA viruses are ideal candidates to induce RNA silencing. A large body of plant viruses have been modified into vectors for RNA silencing in varied plant species. Here, we described a simple, time-saving, and operable system for gene function and genetic breeding study of potato and *Nicotiana benthamiana* using a potato mop-top (MPTV)-based vector.

Key words Potato mop-top virus, Viral vector, dsRNA, RNA-dependent RNA polymerase, Virus-induced gene silencing

1 Introduction

RNA silencing is an evolutionary conserved, double-stranded RNA (dsRNA)-triggered cellular gene expression regulatory mechanism that acts on both transcriptional and posttranscriptional levels [1]. RNA silencing is also an important mechanism that protects against viral infections [2]. The dsRNA form replication intermediates of RNA viruses are ideal candidates to trigger RNA silencing [3]. As a result, many plant viruses, e.g., tobacco rattle virus (TRV), potato virus X (PVX), apple latent spherical virus (ALSV), and bean pod mottle virus (BPMV), have been modified into viral vectors for inducing silencing endogenous genes, a technique namely virus-induced gene silencing (VIGS) [4]. VIGS has been applied to many

Ye Liu and Yanlin Gao contributed equally with all other contributors.

Xiaofei Cheng and Guanwei Wu (eds.), *Double-Stranded RNA: Methods and Protocols*, Methods in Molecular Biology, vol. 2771,
https://doi.org/10.1007/978-1-0716-3702-9_16,

plant species such as *Arabidopsis thaliana*, *Nicotiana benthamiana*, soybean (*Glysine max)*, tomato (*Solanum Lycopersicum*), barley (*Hordei ventilat*), cucumber (*Cucumis sativus*), grape (*Vitis vinifera*), and citrus (*Citrus reticulata*) [5].

Potato mop-top virus (PMTV), the type species of the genus *Pomovirus* in the family *Virgaviridae*, is a destructive plant virus on potato by causing dark arcs and rings on the tuber epidermis [6]. PMTV is transmitted by *Spongospora subterranea* in the field and can also be transmitted mechanically [7]. PMTV has rod-shaped viral particles with tripled RNA genomes termed RNA1, RNA2, and RNA3 based on the size [8]. The genome RNA is capped at the 5′ termini, a transfer RNA (tRNA)-like structure but not the poly A tail was found at the 3′-end [9]. RNA is about 6.1 kilobases (kb) in length, which encodes two replication-associated proteins; RNA2 is about 3.1 kb in size and encodes the two viral coat proteins (CP); RNA 3 is about 2.9 kb and encodes the three partial overlapped open reading frames termed triple-gene-block and a small cysteine-rich protein [10]. Here, we described a step-by-step protocol using an MPTV-based vector to produce viral dsRNA replication intermediates for inducing gene silencing on *Nicotiana benthamiana*.

2 Materials

2.1 Materials

1. Seeds of *N. benthamiana*.
2. The plasmids of PMTV infectious clone, pCB301-PMTVR1, pCB301-PMTVR2, and pCB301-PMTVR3 (*see* **Note 1**).
3. Competent cells of *Escherichia coli* strains DH5α.
4. Competent cells of *Agrobacterium tumefaciens* (synonym *Rhizobium radiobacter*) strain GV3101 harboring the p19 of tomato bushy stunt virus (e.g., IntactGenomics GV3101 pSoup-p19 ElectroCompetent Agrobacterium, Cat no. 1282P19–12) (*see* **Note 2**).
5. Primer premier 5.0 software.
6. Software for DNA sequence analysis (e.g., Lasergene v17) (*see* **Note 3**).

2.2 Reagents and Buffers

All chemicals are analytical grade. Store all buffers at room temperature (unless indicated otherwise).

1. Plant total RNA isolation kit (e.g., Vayzme FastPure Universal Plant Total RNA Isolation Kit, Cat no. RC411-01).
2. 1st strand cDNA synthesis kit (e.g., Vayzme HiScript II 1st Strand cDNA Synthesis Kit with gDNA wiper, Cat no. R212–01).

3. High fidelity DNA polymerase (e.g., Vayzme 2 × Phanta Flash Master Mix Dye Plus, Cat no. P520-01).
4. Kit for DNA fragment recovery from gel (e.g., Vayzme FastPure Gel DNA Extraction Mini Kit, Cat no. DC301-01).
5. Taq DNA polymerase (e.g., Vayzme 2 × Taq Master Mix Dye Plus, Cat no. P112-01).
6. Recombinant DNA clone kit (e.g., Vayzme ClonExpress Ultra One Step Cloning Kit V2, Cat no. C116-01).
7. Plasmid purification kit (e.g., Vayzme FastPure Plasmid Mini Kit, Cat no. DC201-01).
8. Quantitative PCR kit (e.g., Vazyme ChamQ Blue Universal SYBR qPCR Master Mix, Cat no. Q312-02).
9. One-step RT-PCR kit (e.g., Vayzme HiScript II One Step RT-PCR Kit, Cat no. P611-01).
10. Antibiotics: kanamycin (Kana; Merck Sigma Aldrich, Catalog no. E004000), rifampicin (Rif; Merck Sigma Aldrich, Catalog no. 557303), and tetracycline (Tet; Merck Sigma Aldrich, Catalog no. 58346-M) (see **Note 4**).
11. LB medium: 10 g/L tryptone, 10 g/L sodium chloride (NaCl), 5 g/L yeast extract (*see* **Note 5**).
12. Agrobacterium infiltration buffer: 10 mM MgCl2, 10 mM MES, and 150 μM acetosyringone, pH = 5.6.

2.3 Equipment and Software

1. 210 mm garden pots.
2. Soil mixture (e.g., SunGro Sunshine® Mix #1).
3. Plant growth chamber,
4. PCR machine (e.g., LongGene T30).
5. Laboratory shaking incubator (e.g., IKA KS 4000).
6. Pipettor (20, 100, and 1000 μL).
7. Biosafety cabinet.
8. Refrigerated Benchtop centrifuge (e.g., Eppendorf 5418 R Centrifuge).
9. Gel electrophoresis system (e.g., Liuyi Biotechnology Model 122–3220).
10. Electroporation system (e.g., Bio-Rad MicroPulser Electroporator and Electroporation Cuvettes, 0.1 cm gap).
11. Quantitative PCR machine (e.g., Bio-Rad CFX96 touch).

3 Methods

3.1 Determine the Target Fragment and Design Primer

1. Run a blast search using the nucleotide sequence of target gene (e.g., the *N. benthamiana* phytoene desaturase gene, *NbPDS*; Genbank accession no. DQ469932.1) to get all homologs in the genome (*see* **Note 6**).
2. Retrieve the nucleotide homolog sequences of all from GenBank.
3. Multi-align all homologs using the MegAlign in the Lasergene package.
4. Select a conserved fragment as the insertion fragment (e.g., 875–1243 nt) (*see* **Note 7**).
5. Design primer using the Primer premier 5.0 software with default parameters (*see* **Note 8**).

3.2 Construct of pCB301-PMTVR3 Harboring Fragment of Target Gene

1. Extract the total RNA from *N. benthamiana* leaf tissue using the plant total RNA isolation kit according to the user manual.
2. Synthesize the first strand of cDNA using oligo-dT18 primer.
3. Amplify the target fragment by PCR using the primer pair from **step 5** of Subheading 3.1, **step 1** and high-fidelity DNA polymerase.
4. Analyze the PCR results by gel electrophoresis.
5. Recover the fragment from the gel using the Kit for DNA recovery from agarose gel.
6. Linear the pCB301-PMTVR3 plasmid by PCR with high-fidelity DNA polymerase (*see* **Note 9**).
7. Insert the target DNA fragment from **step 5** of Subheading 3.1, **step 2** into linear pCB301-PMTVR3 by recombinant cloning (Fig. 1).
8. Transfer the reaction mix from **step 2** into *E. coli* by heat shocking.
9. Culture the transformant on solid LB medium harboring 50 mg/L Kana.
10. Select the correct colonies by colony PCR with Taq DNA polymerase.
11. Culture the correct colony in liquid LB medium at 37 °C overnight on a shaking incubator.
12. Extract the plasmid from correct colony using plasmid mini extraction kit.
13. Confirm the plasmid by Sanger sequencing.

```
                                                   pPMTVR3-R
GTTGTGGAAAGACGATTTATCGTTTACTATAATTGCAGTGACTGTGCTTGTGGGTGCTATGTT
 V  V  E  R  R  F  I  V  Y  Y  N  C  S  D  C  A  C  G  C  Y  V
                   GmPDS-F
AGCTCGTTATTActgacgagctttcgatgcagtgcattttgattgctttgaacagatttcttc
 S  S  L  L

aggagaaacatggttcaaaaatggccttttagatggtaaccctcctgagagactttgcatgc

cgattgtggaacatattgagtcaaaaggtggccaagtcagactaaactcacgaataaaaaaga

tcgagctgaatgaggatggaagtgtcaaatgttttatactgaataatggcagtacaattaaag

gagatgcttttgtgtttgccactccagtggatatcttgaagcttcttttgcctgaagactgga

                                                    GmPDS-F
aagagatcccatatttccaaaagttggagaagctagtgggagttcctgtgataaatgtccata

         MPTVR3-F
tatACAGTGTTCGGTGCTTGTGTGAATATATGTGTTTGTATTGTATTTTTCATTCTCGTGTAT
    T  V  F  G  A  C  V  N  I  C  V  C  I  V  F  F  I  L  V  Y
```

Fig. 1 Illustration of primer design. Note: The upper and lower lines indicate nucleotide and amino acid sequences, respectively. The CRP sequence is capitalized and the partial sequence of *NbPDS* is in lower case. The primer pair pPMTVR3-F and pPMTVR3-R (pink) are used for amplifying the pCB301-PMTVR3 plasmid, the primer pair NbPDS-F and NbPDS-R are used for amplifying the partial fragment of *NbPDS*

3.3 Infection of N. benthamiana

1. Transfer the pCB301-PMTVR1, pCB301-PMTVR2, and the modified pCB301-PMTVR3-PDS into *Agrobacterium* GV3101 via electroporation.
2. Culture the transformation on solid LB medium containing 50 mg/L Kana, 10 mg/L tet, and 10 mg/L Rif at 28 °C for 2–3 days.
3. Select the correct colonies by colony PCR with Taq DNA polymerase.
4. Culture the correct colony in liquid LB medium containing 50 mg/L Kana, 10 mg/L tet, and 10 mg/L Rif at 28 °C overnight on a shaking incubator.
5. Collect the bacteria by centrifugation at 6000 rpm at room temperature.
6. Resuspend the bacteria in agrobacterium infiltration buffer using a 1 mL pipette tip.

7. Centrifuge to collect the bacteria at room temperature.
8. Repeat **steps 6** and 7 again.
9. Resuspend the bacteria in agrobacterium infiltration buffer using a 1 mL pipette tip.
10. Measure and adjust the O.D. 600 value of bacteria solution to 1.0.
11. Leave the bacteria solution at room temperature for 1–3 h.
12. Mix the bacteria harboring pCB301-PMTVR1, pCB301-PMTVR2, and modified pCB301-PMTVR3-PDS at equal ratio.
13. Inject the mixture into well-expanded *N. benthamiana* leaves using a 1-mL needless syringe.
14. Move the seedlings back to the growth chamber.
15. Check the symptom of PMTV in the systemic leaves (*see* **Note 10**).

3.4 Phenotype Observation and Silencing Efficiency Evaluation

1. Check and record the viral symptoms at one-week post-inoculation and observe every day (*see* **Note 11**).
2. Extract the total RNA from upper systemic soybean leaves using the plant total RNA isolation kit according to the user manual at suitable time point (*see* **Note 12**).
3. Synthesize the first strand of cDNA using oligo-dT18 primer.
4. Evaluation of the silencing efficiency by quantitative PCR (*see* **Note 13**).

3.5 Results

N. benthamiana seedlings that were inoculated by PMTV harboring a fragment of the *NbPDS* gene developed typical bleaching phenotype at about 14 days post inoculation, while those negative control plants are still symptomless (Fig. 2). RT-qPCR is necessary for confirming the silencing efficiency for the subsequent gene function study. The primer NbPDS-qRT-F (5′- CCCGCAGAA GAGTGGATAAA-3′) and NbPDS-qRT-R (5′-TTGGGGTTTTG ACAACATGA-3′) were designed to check the silencing efficiency. RT-qPCR showed that the expression of *GmPDS* was reduced by about 40% in this study (Fig. 3).

4 Notes

1. The PMTV infectious clone used in this protocol was described in [11].
2. The p19 is a strong RNA silencing suppressor, which can significantly improve the expression of PMTV viral RNA [12, 13].

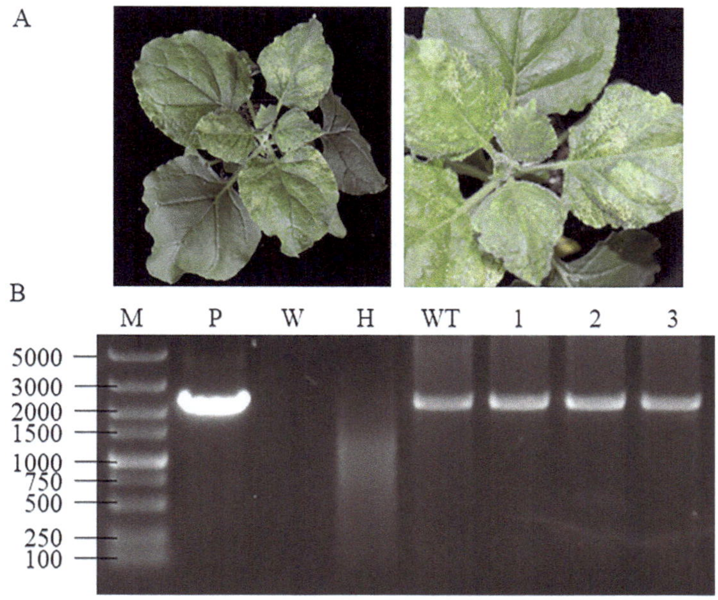

Fig. 2 The phenotype of *N. benthamiana* plant infected by PMTV-PDS. Note: (**a**) phenotype of *N. benthamiana* plants at 14 dpi; (**b**) Confirmation of the presence of PMTV RNA2 in the systemic leaves by RT-PCR; M, P, W, and H, represent DNA marker, plasmid control, mock control, and healthy plant control, respectively; Lanes 1 to 3 represent three biological repeats

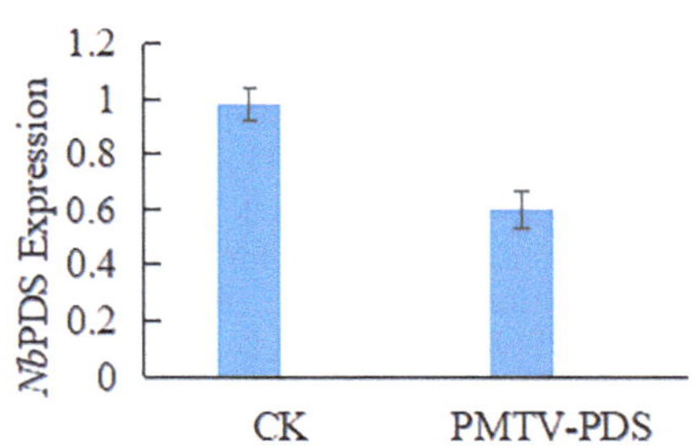

Fig. 3 Evaluation of silence efficiency. Note: Data are the mean of three biological repeats. The expression of *N. benthamiana Actin II* gene was used as the internal control

3. Other DNA analysis software can also be used, e.g., DNAman, Snapgene.
4. Rifampicin should be dissolved in dimethyl sulfoxide (DMSO).
5. Solid LB medium can be prepared by adding 1.2–1.5% agar.
6. Always using the extron sequence for inducing RNA silencing at posttranscriptional level or promoter regions for inducing gene silencing at the transcriptional level (via DNA methylation).
7. An example is provided in Fig. 1.

8. The primers contain 25 bp overlapping sequence to the MPTV RNA3 at each end for recombinant cloning.
9. Since the *CRP* gene is not essential for the survival of PMTV and deletion of *CRP* gene can reduce viral symptom severity [14]. The exogenous DNA fragment was inserted into the *CRP* gene. The primers used for PCR are shown in Fig. 1.
10. One primer should be located in the PMTV RNA2 and another located inside the fragment to avoid amplifying endogenous gene. An example is provided in Fig. 1.
11. A positive control experiment with PMTV harboring a visible marker gene such as *GmPDS* should be parallelly performed for studying endogenous genes.
12. MPTV typically takes about 12–14 days to infect the systemic leaves.
13. *NbActin* is a widely used internal control; however, the *eIF1a* is also an excellent candidate as the internal control.

References

1. Baulcombe D (2004) RNA silencing in plants. Nature 431(7006):356–363
2. Plasterk RHA (2002) RNA silencing: the genome's immune system. Science 296(5571): 1263–1265
3. Carrington JC, Kasschau KD, Johansen LK (2001) Activation and suppression of RNA silencing by plant viruses. Virology 281(1):1–5
4. Purkayastha A, Dasgupta I (2009) Virus-induced gene silencing: a versatile tool for discovery of gene functions in plants. Plant Physiol Biochem 47(11–12):967–976
5. Lange M, Yellina AL, Orashakova S, Becker A (2013) Virus-induced gene silencing (VIGS) in plants: an overview of target species and the virus-derived vector systems. In: Becker A (ed) Virus-induced gene silencing: methods and protocols. Humana Press, Totowa, pp 1–14
6. Adams MJ et al (2017) ICTV virus taxonomy profile: virgaviridae. J Gen Virol 98(8):1999
7. Savenkov EI, Germundsson A, Zamyatnin AA, Sandgren M, Valkonen JPT (2003) Potato mop-top virus: the coat protein-encoding RNA and the gene for cysteine-rich protein are dispensable for systemic virus movement in *Nicotiana benthamiana*. J Gen Virol 84(4): 1001–1005
8. Hu X et al (2016) Molecular characterization of potato mop-top virus isolates from China and Canada and development of RT-PCR differentiation of two sequence variant groups. Can J Plant Pathol 38(2):231
9. Savenkov EI, Sandgren M, Valkonen JPT (1999) Complete sequence of RNA 1 and the presence of tRNA-like structures in all RNAs of potato mop-top virus, genus Pomovirus. J Gen Virol 80(Pt 10):2779–2784
10. Lukhovitskaya NI et al (2005) Expression, localization and effects on virulence of the cysteine-rich 8 kDa protein of potato mop-top virus. J Gen Virol 86(Pt 10): 2879–2889
11. Liu Y et al (2022) Cloning of potato mop-top virus genome and construction of infectious cDNA clone. China Potato 36(4):322–331
12. Vargason JM, Szittya G, Burgyan J, Hall TM (2003) Size selective recognition of siRNA by an RNA silencing suppressor. Cell 115(7): 799–811
13. Ye K, Malinina L, Patel DJ (2003) Recognition of small interfering RNA by a viral suppressor of RNA silencing. Nature 426(6968):874–878
14. Kalyandurg PB et al (2019) Efficient RNA silencing suppression activity of potato mop-top virus 8K protein is driven by variability and positive selection. Virology 535:111–121

Chapter 17

Application of dsRNA for Fungi Disease Management *Sclerotinia sclerotiorum* and *Botrytis cinerea*

Tao Tang, Yumeng Wang, and Dongdong Niu

Abstract

Sclerotinia sclerotiorum and *Botrytis cinerea* are two closely related necrotrophic plant pathogenic fungi that have garnered significant attention due to their broad host range and persistent environmental presence. The use of double-stranded RNA (dsRNA) to inhibit specific genes has demonstrated promising potential in disease management. Various RNA interference (RNAi) strategies have been developed and applied in disease management, including the use of dsRNA to silence specific genes, as well as other RNA-based approaches. These strategies show promise in controlling fungal diseases and have the potential for broader application in plant protection. Here, we present a protocol that outlines techniques for preparing dsRNA and spraying it directly to control fungal diseases. This protocol offers a practical and effective solution for managing plant diseases and has the potential to be widely applied in disease management.

Key words dsRNA, RNA interference, Spray-induced gene silencing

1 Introduction

S. sclerotiorum and *B. cinerea*, both necrotrophic plant pathogenic fungi with a wide host range, have a close relationship in the phylogenetic tree, and each causes millions of dollars in crop losses worldwide [1, 2]. Currently, the prevention and control of these diseases mainly rely on chemical fungicides. However, the overuse of fungicides can lead to environmental problems. RNA silencing technology (RNAi) offers promising possibilities for disease control [3]. Multiple RNAi strategies have been applied to disease prevention and control, achieving remarkable results [4–7]. Among these methods, direct spraying of dsRNA for the control of plant diseases displays particular advantages. This method does not require GMO operation, thus shortening the duration of the experiment. Additionally, the easily degradable nature of dsRNA results in almost no residue on crops.

Xiaofei Cheng and Guanwei Wu (eds.), *Double-Stranded RNA: Methods and Protocols*, Methods in Molecular Biology, vol. 2771,
https://doi.org/10.1007/978-1-0716-3702-9_17,

The application of dsRNA requires the synthesis of dsRNA, spraying dsRNA on plants, and phenotypic analysis of the control effect. Here, we provide a complete experimental protocol.

1.1 Considerations for the dsRNA Design

Through preliminary experiments, pathogenic genes or essential growth genes of pathogenic bacteria were identified. When selecting a DNA sequence of a specific gene as a template for synthetic dsRNA, consider the following points:

1. Region: The template DNA sequence should be derived from exons rather than introns. To obtain the correct coding sequence of genes, compare transcriptome and genomic data to obtain a cDNA clone. Additionally, obtain a CDS clone containing a contiguous stretch of exonic sequence through RT-PCR, which can be used as the template.
2. Length: Longer dsRNAs may be more effective in triggering RNAi because they can target more mRNA regions and be sliced into more siRNA molecules than shorter dsRNAs. Considering longer sense and anti-sense strands RNA may be more challenging to anneal. The recommended length of dsRNA is about 200 bp to 300 bp, and it should be shorter than 800 bp.
3. Specificity: The dsRNA sequence should be specific to the target gene to avoid off-target effects. To ensure specificity, perform BLAST analysis against the entire genome to identify potential off-target sequences.

1.2 Considerations for dsRNA Application Method

The concentration of dsRNA applied can impact its efficacy in disease control. Higher concentrations may offer better protection, but they can also be more expensive and potentially harmful to non-target organisms. Direct spraying of dsRNA on the plant surface can provide effective protection. However, dispersing the dsRNA solution on certain host plants with surface barriers, such as tomato fruit with a waxy layer or mature rice leaves with silicified cells, can be challenging. To aid in dispersal, an appropriate amount of Tween can be added to the dsRNA solution. dsRNA degrades rapidly in the environment, providing it with a unique advantage that not easily generate residues while limiting its long-term protection of crops. To enhance its protective effect and prolong the duration of protection, dsRNA is often loaded onto nanoparticles with negatively charged surfaces. At present, the widely used nanomaterials include layered bilayer hydroxides (LDH) [8], carbon dots (CDs) [9], and others.

2 Materials

1. PDA medium: Potato dextrose broth (24 g) and 15 g of agar per liter.
2. Malt extract agar medium and sabouraud maltose broth: Malt extract (20 g), bacto protease peptone(10 g), and extra add 15 g of agar per liter for malt extract agar medium.
3. DNA templates for dsRNA synthesis.
4. dsRNA prepared by in vitro transcription.

3 Methods

3.1 S. sclerotiorum and B. cinerea Culture

S. sclerotiorum strain was routinely cultured on PDA medium and *B. cinerea* was cultured on malt extract agar medium. For *B. cinerea* infection, the fungal spores were diluted in sabouraud maltose broth.

3.2 Acquisition of dsRNA/Synthesis of dsRNA by In Vitro Transcription

3.2.1 DNA Templates for dsRNA Synthesis

DNA templates could be generated by PCR reaction; the primers used in this reaction have T7 promoters in their 5′ ends so that the PCR fragments could be flanked by T7 promoters. The PCR reaction can be performed on genomic DNA directly from fungus, or on plasmid containing the CDS of target gene and the latter protocol is preferred because it is less prone to non-specific amplification. Before synthesizing dsRNA, the PCR template can be purified by gel purification to reduce non-specific products (a number of commercial kits are available). The schematic for dsRNA synthesis is shown in Fig. 1.

To generate a template for dsRNA synthesis, design a suitable PCR program based on the selected PCR polymerase and primers. After PCR amplification, analyze the PCR products by agarose gel electrophoresis to confirm that the products contain the required DNA fragments. The enzymes and salts in PCR products could be removed by purifying over a DNA affinity column (commercial kits are available) or by utilizing phenol/chloroform extraction and ethanol precipitation.

3.2.2 In Vitro Transcription dsRNA

The following protocols describe how dsRNA is produced from such templates.

1. Add the following to an EP tube on ice:
   - (a) 1 μg PCR product (100–1000 ng).
   - (b) 2 μL of 5 mM NTPs.
   - (c) 2 μL of 10X transcription buffer (supplied with the RNA polymerases: 400 mM Tris-HCl; 60 mM MgCl, 100 mM dithiotreitol (DTT), 20 mM spermidin).

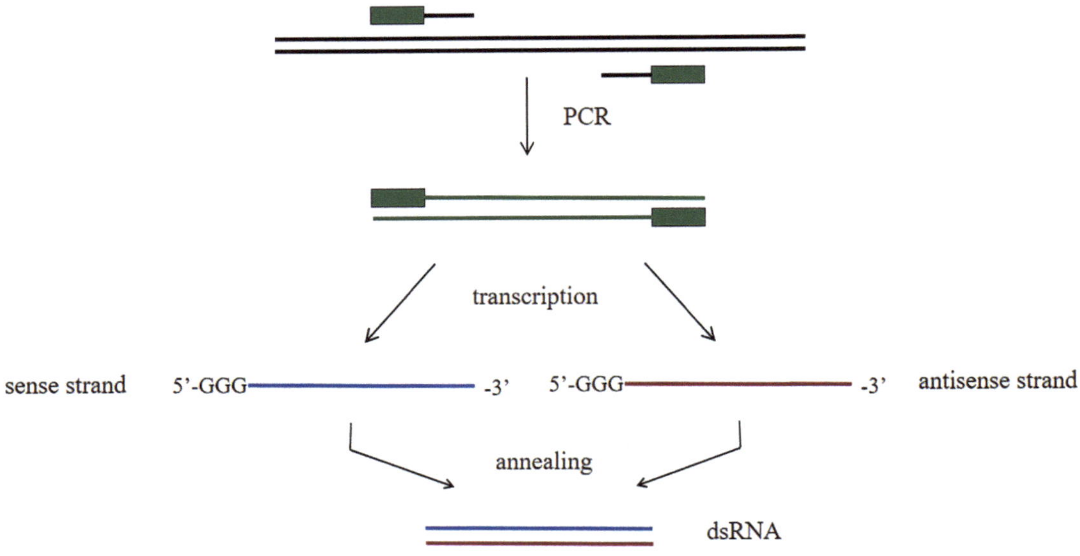

Fig. 1 Schematic diagram of the in vitro transcriptional process of dsRNA. The green horizontal line in the diagram is the T7 promoter sequence

(d) 15–20 U of RNA polymerase (stored on ice).

(e) Add RNase-free double distilled water to a final volume of 20 μL.

2. Mix and centrifuge briefly and incubate for 2 h at 37 °C.
3. Add 2 μL DNase I and incubate for 15 min at 37 °C to remove template DNA. Stop the reaction by adding 2 μL 0.2 M EDTA.
4. Purification of RNA products by adsorption columns or magnetic beads.
5. Agarose gel electrophoresis is used to detect the synthesized dsRNA (Fig. 2).

3.3 Application of dsRNA in Plants to Manage Fungal Disease

The selection of appropriate target genes is critical for the successful application of dsRNA in managing fungal diseases. Preliminary laboratory studies have demonstrated that the Dicer genes of *B. cinerea* and *S. sclerotiorum* are essential for their pathogenesis, making them suitable candidates for dsRNA targeting. Before application, dsRNAs are prepared at the appropriate concentration in RNase-free water. In vitro synthesized dsRNAs are diluted to 40 ng/μL, while prokaryotic expressed dsRNAs, which are of lower purity, are used at a concentration of 1000 ng/μL. The dsRNA is sprayed onto the plant surface, and the pathogenic fungus is inoculated at the same site.

For *B. cinerea* infections, fungal spores are diluted in 1% Sabouraud maltose broth buffer to a final concentration of 10^5 spores/mL and drip-inoculated onto the plants. For

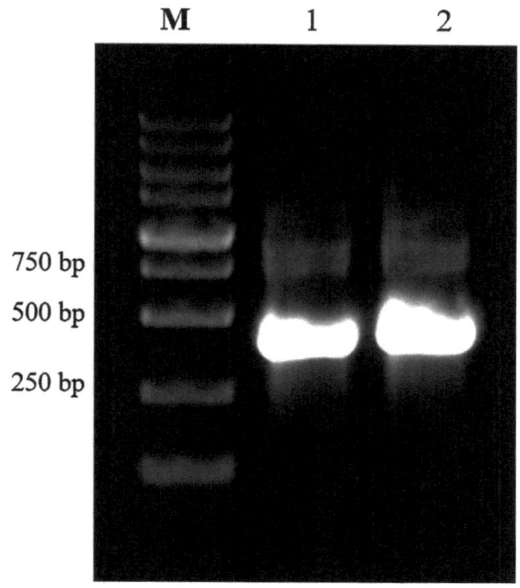

Fig. 2 Agarose gel electrophoresis of in vitro transcribed dsRNA. M is a 5000 bp DNA marker; Lane 1 is *BcDCL1/2* -dsRNA at 490 bp and lane 2 is *SsDCL1/2* -dsRNA at 460 bp

S. sclerotiorum infections, mycelial plugs (5 mm in diameter) obtained from the margins of newly activated colonies are used to inoculate vegetable leaves. Infected leaf tissue is then incubated at 25 °C in a light incubator.

4 Notes

In the preceding section, we have outlined the standard protocol for dsRNA production. In the subsequent section, our objective is to comprehensively elucidate the critical factor essential that must be taken into account in the production, purification, and application of dsRNA. This endeavor seeks to provide readers with a deep understanding of these pivotal considerations, which are imperative to guarantee the purity and efficacy of dsRNA for a diverse spectrum of applications.

It is paramount to underscore that the quality of the DNA template employed for in vitro transcription and dsRNA synthesis bears a profound impact on the efficiency of in vitro transcription. Attaining high-quality DNA templates necessitates meticulous attention to the PCR reaction's efficiency and specificity. This is achieved through the utilization of high-quality, high-quantity PCR templates, such as plasmids housing target genes. Furthermore, the judicious selection of the appropriate PCR polymerase and primers is imperative to secure the desired PCR product. Additionally, the practice of gel purification is strongly recommended to eliminate any non-specific byproducts that may impede subsequent stages of dsRNA synthesis. During primer design, the T7 promoter sequence should be added to the 5′ end of the primer, with the T7 promoter sequence "taatacgactcactataggg" precisely corresponding to the transcription start site "GGG."

Reagents used in the in vitro transcription process, such as DNA templates and RNA polymerases, should be of high quality and free of RNAase contaminants to ensure optimal yield and purity. Optimizing the RNA polymerase concentration and reaction conditions, such as pH, temperature, and buffer composition, can also contribute to maximizing the yield and purity of synthesized dsRNA. After transcription, it is essential to purify the RNA to remove any residual DNA, enzymes, or impurities that may affect downstream applications. Verifying the quality and quantity of the synthesized dsRNA through techniques such as gel electrophoresis, UV spectrophotometry, and RNA sequencing can help ensure that the dsRNA is of sufficient quality and purity for subsequent experiments.

In the process of applying dsRNA to plants for disease control, it is important to note that for certain plants with hydrophobic surfaces, achieving an even dispersion of dsRNA across the plant surface can be facilitated by adding an appropriate quantity of Tween to the dsRNA solution.

References

1. Bolton MD, Thomma BP, Nelson BD (2006) Sclerotinia sclerotiorum (lib.) de Bary: biology and molecular traits of a cosmopolitan pathogen. Mol Plant Pathol 7:1–16
2. Williamson B, Tudzynski B, Tudzynski P, van Kan JAL (2007) *Botrytis cinerea*: the cause of grey mould disease. Mol Plant Pathol 8:561–580
3. Wytinck N, Manchur CL, Li VH et al (2020) dsRNA uptake in plant pests and pathogens: insights into RNAi-based insect and fungal control technology[J]. Plan Theory 9(12):1780
4. Koch A, Kumar N, Weber L et al (2013) Host-induced gene silencing of cytochrome P450 lanosterol C14α-demethylase–encoding genes confers strong resistance to Fusarium species [J]. Proc Natl Acad Sci 110(48):19324–19329
5. Koch A, Biedenkopf D, Furch A et al (2016) An RNAi-based control of Fusarium graminearum infections through spraying of long dsRNAs involves a plant passage and is controlled by the fungal silencing machinery [J]. PLoS Pathog 12(10):e1005901
6. Qi T, Guo J, Peng H et al (2019) Host-induced gene silencing: a powerful strategy to control diseases of wheat and barley [J]. Int J Mol Sci 20(1):206
7. Qiao L, Lan C, Capriotti L et al (2021) Spray-induced gene silencing for disease control is dependent on the efficiency of pathogen RNA uptake [J]. Plant Biotechnol J 19(9):1756
8. Mitter N, Worrall EA, Robinson KE et al (2017) Clay nanosheets for topical delivery of RNAi for sustained protection against plant viruses[J]. Nat Plants 3(2):1–10
9. Wang Z, Li Y, Zhang B et al (2023) Functionalized carbon dot-delivered RNA Nano fungicides as superior tools to control *Phytophthora* pathogens through plant *RdRP1* mediated spray-induced gene silencing[J]. Adv Funct Mater 2213143

Chapter 18

Application of dsRNA in the Pine Wood Nematode, *Bursaphelenchus xylophilus*

Chunyu Wang and Kai Guo

Abstract

The pine wood nematode *Bursaphelenchus xylophilus* is one of the most destructive invasive species worldwide, causing the wilting and eventual death of pine trees. Despite recognition of their economic and environmental significance, it has thus far been impossible to study the detailed gene functions of plant parasitic nematodes through conventional forward genetics and transgenic methods. RNA interference (RNAi), as a reverse genetics technology, offers great convenience for studying the functional genes of nematodes, including *B. xylophilus*. We here outline a protocol for RNAi of the *ppm-1* gene in *B. xylophilus*, which has been reported to play crucial roles in the development and reproduction of other pathogenic nematodes. For RNAi, the T7 promoter was linked to the 5′-terminal of the target fragment by polymerase chain reaction (PCR), and then double-stranded RNA (dsRNA) was synthesized by in vitro transcription. Subsequently, dsRNA delivery was accomplished by soaking nematodes with the dsRNA solution mixed with synthetic neurostimulants. Synchronized eggs, juveniles, and adults of *B. xylophilus* (approximately 20,000 individuals of each stage) were washed and soaked in dsRNA (0.8 μg/μL) with the soaking buffer for 24 h in the dark at 25 °C. The same quantity of nematodes was placed in the soaking buffer without dsRNA or with green fluorescent protein dsRNA as a control. After soaking, the expression level of the target transcripts was determined by real-time quantitative PCR. The effects of RNAi were then confirmed by microscopic observation of the phenotypes and comparison of the body size of adults among groups. The current protocol can help to progress research to understand the functions of the genes of *B. xylophilus* and other parasitic nematodes toward developing control strategies through genetic engineering.

Key words dsRNA, RNAi, *Bursaphelenchus xylophilus*

1 Introduction

Bursaphelenchus xylophilus is the causal pathogen of pine wilt disease, which has caused great harm to pine forests worldwide [1]. To date, there has been substantial research on pine wilt disease with most of these studies focused on exploring the infection mechanism

The original version of the chapter has been revised. A correction to this chapter can be found at https://doi.org/10.1007/978-1-0716-3702-9_19

Xiaofei Cheng and Guanwei Wu (eds.), *Double-Stranded RNA: Methods and Protocols*, Methods in Molecular Biology, vol. 2771, https://doi.org/10.1007/978-1-0716-3702-9_18,

Corrected Publication 2026

and the interactions of pathogenic nematodes, host pines, vector insects, and environmental factors; however, research on the developmental and reproductive biology of *B. xylophilus* remains rare.

RNA interference (RNAi) is a mechanism widely present in eukaryotic cells to regulate gene expression; RNAi silences gene expression by introducing double-stranded RNA (dsRNA) [2]. Since discovery of the RNAi machinery in *Caenorhabditis elegans* in 1998 [3], it has been exploited and developed as a useful experimental technique applicable to many organisms, including nematodes [4]. RNAi techniques have now become effective methods for identifying the gene function of nematodes, and RNAi has also been proposed as a new way to effectively control pathogenic nematodes [5].

Here, we describe a protocol for applying RNAi to study the function of the *ppm-1* gene of *B. xylophilus*, including synthesis of dsRNA, soaking, and quantitative polymerase chain reaction (qPCR) detection. This protocol can also serve as a reference for gene function analyses of other plant parasitic nematodes.

2 Materials

2.1 Nematode Strain and Culture Conditions

The nematode population, *B. xylophilus* NXY61, was originally isolated from diseased *Pinus massoniana* in the Ningbo area, Zhejiang province, China. The fungal food source, *Botrytis cinerea*, was obtained from the Research Institute of Forest Protection, Chinese Academy of Forestry, Beijing, China. Nematodes were cultured in the dark on a lawn of *B. cinerea* in Petri dishes with PDA media at 25 °C.

2.2 Acquisition of Nematodes at Different Development Stages

B. xylophilus strain was cultured on the mycelia of *B. cinerea* in PDA plate at 25 °C for 3–5 days. Keep nematodes in the glass Petri dish for 30 min, the eggs will stick to the bottom of the glass Petri dish. Remove the water and nematodes carefully, without disturbing the eggs and repeat the steps until remove all the larvae and adults, and obtain the synchronized eggs. Hatch the collected eggs for 24 h in the dark at 25 °C to obtain J2 larvae. Synchronized J2 was cultured on *B. cinerea* culture on PDA plate at 25 °C for 24, 48, and 72 h, and the synchronized J3, J4 and adult nematodes were obtained respectively. The synchronized juveniles, J3, J4 and adult nematodes were extracted by the Baermann funnel method, respectively.

3 Methods

3.1 Synthesis of dsRNA

1. Extract total RNA from nematodes using TRIzol reagent.
2. Clone *ppm-1* gene sequences into the vector plasmid contain T7 promoter following a standard cloning protocol.

3. Design specific primers according to the sequence of the *ppm-1* gene (*ppm-1* RNAi T7 F: 5′- TAATACGACTCACTATA GGGGATGCGAAGTTGCCAATCATTCT -3′; R: 5′- TAA TACGACTCACTATAGGGCCAGATCCAGTCCACCATA CACC-3′).
4. Use the plasmid containing the *ppm-1* gene fragment as the template for PCR and then recover the fragment containing the T7 promoter.
5. Use T7 Transcription Kit to synthesize dsRNA in vitro.
   (a) The reaction system is as follows:
       ATP solution 2.0 μL.
       CTP solution 2.0 μL.
       GTP solution 2.0 μL.
       UTP solution 2.0 μL.
       10X Reaction Buffer 2.0 μL
       Enzyme Mix 2.0 μL.
       DNA Template 0.1 ~ 1 μg.
       Nuclease-free Water up to 20.0 μL.
       After thorough mixing of the above reagents, incubate at 37 °C for 2–4 h.
   (b) In the above mixture, add 1 μL of TURBO DNase to remove DNA. After thorough mixing, incubate at 37 °C for 15 min.
   (c) Precipitate dsRNA with LiCl:
       Add 30 μL of LiCl and 30 μL of Nuclease-free Water to the mixture, precipitate RNA, and mix well. Transfer the mixture to a 1.5 mL RNase-free centrifuge tube and cool at −20 °C for more than 30 min, or overnight.
   (d) Centrifuge at 4 °C, 12,000 rpm, for 15 min. Carefully remove the supernatant.
   (e) Wash the RNA once with 1 ml of 70% ethanol. Centrifuge at 4 °C, 12,000 rpm, for 10 min. Carefully remove the 70% ethanol and let the remaining ethanol evaporate for 2–3 min with the tube open.
   (f) Suspend the RNA in 25 μL of Nuclease-free water. Measure the RNA concentration using a NanoDrop spectrophotometer. Store at −20 °C.
6. Analyze the quality of the dsRNA using a spectrophotometer and visualize the products on a 1.0% agarose gel.

3.2 RNAi by Soaking

1. In a 200 μL PCR tube, mix 4 μL of 5× soaking buffer with the dsRNA and ddH_2O, resulting in a total volume of 20 μL and a final RNA concentration of 0.8 μg/μL (or 800 ng/μL) (*see* **Note 1**).
2. Wash the synchronous nematodes three times in 0.2 mM M9 buffer (*see* **Note 2**).

3. Transfer the nematodes in the dsRNA solution, and incubate with centrifugation at 15× *g* on a shaking table for 24 h at 25 °C to ensure that the nematodes will effectively absorb the dsRNA [6] (*see* **Note 3**).
4. Soak the same quantity of nematodes in the soaking buffer without the dsRNA probe or with a green fluorescent protein (GFP) dsRNA probe as a control (*see* **Note 4**).

3.3 qPCR Detection

1. Clean the nematodes with ddH_2O, including those with target gene interference, GFP gene interference, and the undisturbed control group. Extract the total RNAs from each group.
2. Start PCR. Use *actb* and *tbb-2* (or other genes) as internal reference genes to evaluate the changes of gene expression level after RNAi.
3. According to the cycle threshold (Ct) value and dissolution curve, use the $2^{-\Delta\Delta Ct}$ method to estimate the relative expression level of the target gene and verify the interference efficiency.

3.4 Evaluate the Body Length of Adults After RNAi

1. After RNAi to J2-stage nematodes, collect and transfer the mycelia of *B. cinerea* to potato dextrose agar-containing Petri dishes for 60 h at 25 °C for development into adults [7] (*see* **Note 5**).
2. Collect the adults by the Bellman funnel method [8].
3. Acquire images of the adult nematodes under a microscope and use ImageJ software (or other measurement software) to measure the body lengths; measurements of 50 male and 50 female nematodes should be taken for statistical analysis [9].
4. Analyze the data by calculating the mean and standard deviation for each sample. The means of samples from different groups can be compared using Student's t-test (*see* **Note 6**).

3.5 Representative Results

3.5.1 Analysis of ppm-1 Expression of B. xylophilus After RNAi

The relative expression level of the *ppm-1* gene of *B. xylophilus* soaked with GFP dsRNA and soaked with target gene dsRNA was 0.92 and 0.52, respectively (the *ppm-1* gene expression level of the ddH_2O-treated control group was set to 1) (Fig. 1). Thus, exogenous dsRNA has no effect on the *ppm-1* expression of *B. xylophilus*, but *ppm-1* dsRNA can effectively inhibit the expression of the target gene.

3.5.2 Effect of ppm-1 Expression on Growth and Development of B. xylophilus

After RNAi, the size of the adults markedly decreased (Fig. 2), resulting in the SMA (small body size) mutant phenotype. Although RNAi-treated individuals developed to sexual maturity, the body length was substantially smaller than that of normal adults. Specifically, after RNAi to J2-stage nematodes, the mean

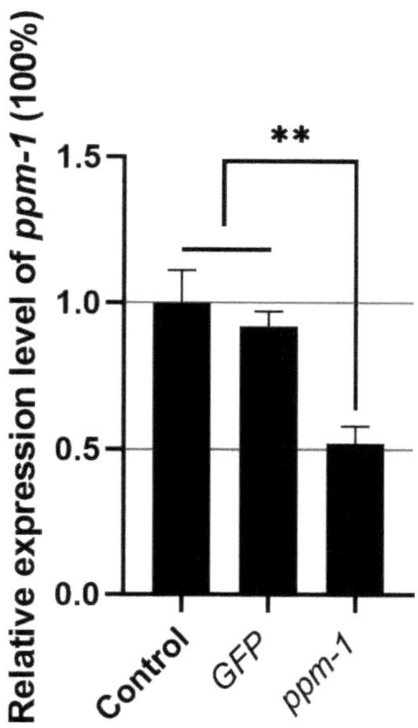

Fig. 1 Expression of the *ppm-1* gene following RNAi of *B. xylophilus*. $**P < 0.001$

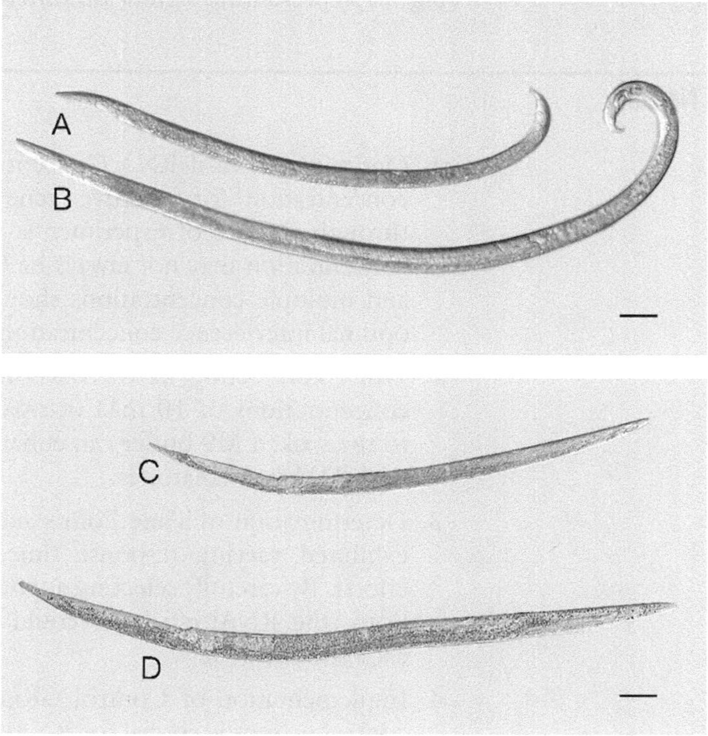

Fig. 2 Body length reduction of adults after interference of *ppm-1* in *B. xylophilus*. Images of an adult male (A) and female (C) of the RNAi group. Images of an adult male (B) and female (D) of the control group. Scale bar: 50 μm

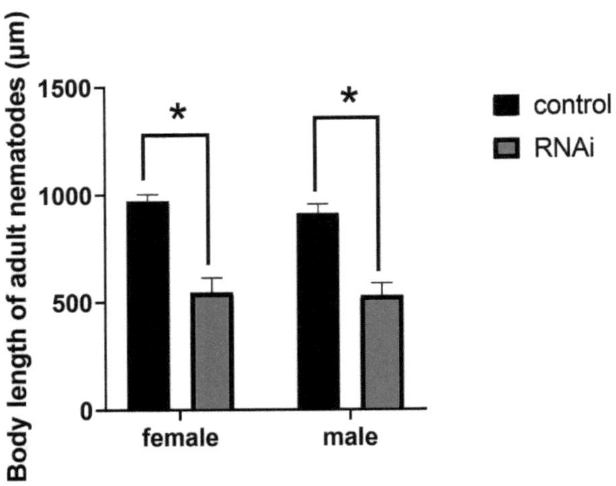

Fig. 3 Quantification and statistical analysis of the body length of adults of *B. xylophilus* after RNAi of the *ppm-1* gene. $*P = 0.0322$

body length of females and males was 544.61 μm and 526.24 μm, respectively, whereas the mean body length of females and males in the control group was 971.86 μm and 912.31 μm, respectively (Fig. 3), representing significant differences ($P = 0.0322$).

4 Notes

1. Optimization of dsRNA Concentration: The optimal dsRNA concentration for effective gene silencing was determined through a series of experiments. It was found that the ideal concentration may not always be 0.8 μg/μL (or 800 ng/μL), and multiple concentrations should be tested to identify the optimal interference concentration.
2. When conducting in vitro RNAi experiments, adding final concentrations of 10 mM octopamine or 1% phenylthiourea to the soaked M9 buffer can enhance the absorption efficiency of dsRNA by nematodes.
3. Determination of Time Points and Duration: Different genes exhibited varying response times and durations for RNAi effects. By carefully selecting appropriate time points and durations, the RNAi efficiency could be accurately evaluated for each specific gene.
4. Implementation of Control Groups: The use of appropriate control groups is crucial for RNAi experiments. Negative control groups, such as non-specific RNAi probes or GFP RNAi probes, were employed to account for nonspecific effects. Additionally, a positive control group comprising genes known to be inhibited by RNAi was included to validate the effectiveness of RNAi.

5. Immediately after soaking, nematodes may exhibit a rigid state. It is necessary to transfer the nematodes to RNase-free water for recovery for 30 min before conducting morphological observations.
6. Replication of Experiments: To ensure the reliability and reproducibility of the results, experiments were repeated multiple times. The replication of experiments reduced random errors and strengthened the confidence in the obtained results.

Acknowledgments

This research was funded by the National Natural Science Foundation of China (31200487, 31870637) and jointly funded by Zhejiang Key Research Plan (2016C32016, 2019C02024).

Figures and text in this chapter are adapted from Liu et al. (2022) with permission from JoVE.

References

1. Kikuchi T, Cotton JA, Dalzell JJ, Hasegawa K, Kanzaki N, McVeigh P et al (2011) Genomic insights into the origin of parasitism in the emerging plant pathogen *Bursaphelenchus xylophilus*. PLoS Pathog 7(9):e1002219
2. Niu JH, Jian H, Xu JM, Guo YD, Liu Q (2010) RNAi technology extends its reach: engineering plant resistance against harmful eukaryotes. Afr J Biotechnol 9(45):7573–7582
3. Fire A, Xu S, Montgomery MK, Kostas SA, Driver SE, Mello CC (1998) Potent and specific genetic interference by double-stranded RNA in *Caenorhabditis elegans*. Nature 91(6669): 806–811
4. Wang M, Wang D, Zhang X, Wang X, Liu W, Hou X et al (2016) Double-stranded RNA-mediated interference of dumpy genes in Bursaphelenchus xylophilus by feeding on filamentous fungal transformants. Int J Parasitol 46(5–6): 351–360
5. Shahid M, Imran A, Mazhar H, Yusuf Z, Rob WB (2006) Engineering novel traits in plants through RNA interference. Trends Plant Sci 11(11):559–565
6. Dai S (2007) Study on the molecular ecological mechanism of the adaptation of *Bursaphelenchus xylophilus* to temperature stress. Doctoral thesis, Hunan Agricultural University. (2007). https://kns.cnki.net/KCMS/detail/detail.aspx?dbname=CDFD9908&filename=2007208175.nh
7. Zhu N, Bai L, Schütz S, Liu B, Liu Z, Zhang X et al (2016) Observation and quantification of mating behavior in the pinewood nematode, *Bursaphelenchus xylophilus*. J Vis Exp 118:54842
8. Zhou J (2013) Dictionary of soil science. Science Press.(book)
9. Zhou L, Ma X, Zhu N, Zou Q, Guo K, Bai L et al (2020) The role of mab-3 in spermatogenesis and ontogenesis of pinewood nematode, Bursaphelenchus xylophilus. Pest Manag Sci 77(1):138–147

Correction to: Application of dsRNA in the Pine Wood Nematode, *Bursaphelenchus xylophilus*

Chunyu Wang and Kai Guo

Correction to:
Chapter 18 in: Xiaofei Cheng and Guanwei Wu (eds.), *Double-Stranded RNA: Methods and Protocols*, Methods in Molecular Biology, vol. 2771,
https://doi.org/10.1007/978-1-0716-3702-9_18

In Abstract, line 12 on page 133, the text "soaked in dsRNA (0.8 μg/mL) with" has been updated as "soaked in dsRNA (0.8 μg/μL) with". In Section 3.2, on the last line of note 1 in page 135, the text "…a final RNA concentration of 0.8 μg/μL (or 1000 ng/μL) (see Note 1)." has been updated as "…a final RNA concentration of 0.8 μg/μL (or 800 ng/μL) (see Note 1)." In Section 4, page 138, on the 3rd line of the note 1, the text "…ideal concentration may not always be 1000 ng/μL, and multiple…" has been updated as "… ideal concentration may not always be 0.8 μg/μL (or 800 ng/μL), and multiple…"

Additionally, acknowledgments have been updated as follows:

Acknowledgments
This research was funded by the National Natural Science Foundation of China (31200487, 31870637) and jointly funded by Zhejiang Key Research Plan (2016C32016, 2019C02024).

Figures and text in this chapter are adapted from Liu et al. (2022) with permission from JoVE.

The updated version of this chapter can be found at
https://doi.org/10.1007/978-1-0716-3702-9_18

Xiaofei Cheng and Guanwei Wu (eds.), *Double-Stranded RNA: Methods and Protocols*, Methods in Molecular Biology, vol. 2771,
https://doi.org/10.1007/978-1-0716-3702-9_19,

INDEX

Xiaofei Cheng and Guanwei Wu (eds.), *Double-Stranded RNA: Methods and Protocols*, Methods in Molecular Biology, vol. 2771, https://doi.org/10.1007/978-1-0716-3702-9,

MIX
Papier aus verantwortungsvollen Quellen
Paper from responsible sources
FSC® C105338

If you have any concerns about our products,
you can contact us on
ProductSafety@springernature.com

In case Publisher is established outside the EU,
the EU authorized representative is:
Springer Nature Customer Service Center GmbH
Europaplatz 3, 69115 Heidelberg, Germany

Printed by Libri Plureos GmbH
in Hamburg, Germany